Mathematical
Surveys
and
Monographs

Volume 197

The Octogonal PETs

Richard Evan Schwartz

American Mathematical Society
Providence, Rhode Island

2010 *Mathematics Subject Classification.* Primary 37E20, 37E05, 37E15.

For additional information and updates on this book, visit
www.ams.org/bookpages/surv-197

Library of Congress Cataloging-in-Publication Data

Schwartz, Richard Evan.
The octogonal PETs/Richard Evan Schwartz.
pages cm. – (Mathematical surveys and monographs ; volume 197)
Includes bibliographical references.
ISBN 978-1-4704-1522-8 (alk. paper)
1. Polytopes. 2. Geometry. I. Title. II. Title: Octogonal polytope exchange transformations.

QA691.S45 2014
515′.48–dc23 2014006823

∞ The paper used in this book is acid-free and falls within the guidelines
established to ensure permanence and durability.
Visit the AMS home page at `http://www.ams.org/`

10 9 8 7 6 5 4 3 2 1 19 18 17 16 15 14

Contents

Preface ix

Chapter 1. Introduction 1
1.1. What is a PET? 1
1.2. Some Examples 1
1.3. Goals of the Monograph 2
1.4. The Octagonal PETs 3
1.5. The Main Theorem: Renormalization 4
1.6. Corollaries of The Main Theorem 6
1.7. Polygonal Outer Billiards 10
1.8. The Alternating Grid System 11
1.9. Computer Assists 13
1.10. Organization 13

Part 1. Friends of the Octagonal PETs 15

Chapter 2. Background 17
2.1. Lattices and Fundamental Domains 17
2.2. Hyperplanes 17
2.3. The PET Category 18
2.4. Periodic Tiles for PETs 19
2.5. The Limit Set 20
2.6. Some Hyperbolic Geometry 21
2.7. Continued Fractions 23
2.8. Some Analysis 24

Chapter 3. Multigraph PETs 27
3.1. The Abstract Construction 27
3.2. The Reflection Lemma 28
3.3. Constructing Multigraph PETs 29
3.4. Planar Examples 30
3.5. Three Dimensional Examples 30
3.6. Higher Dimensional Generalizations 31

Chapter 4. The Alternating Grid System 33
4.1. Basic Definitions 33
4.2. Compactifying the Generators 34
4.3. The PET Structure 36
4.4. Characterizing the PET 37
4.5. A More Symmetric Picture 38

4.6. Unbounded Orbits 40
4.7. The Complex Octagonal PETs 40

Chapter 5. Outer Billiards on Semiregular Octagons 43
5.1. The Basic Sets 43
5.2. The Far Partition 44
5.3. The First Return Map 45
5.4. The Necklace Orbits 47
5.5. Parallelograms, Halfbones, and Dogbones 47
5.6. The Dogbone Map 49
5.7. The First Conjugacy 50
5.8. The Second Conjugacy 54

Chapter 6. Quarter Turn Compositions 55
6.1. Basic Definitions 55
6.2. The Polytope Graph 55
6.3. QTCs and Polygon Graphs 57
6.4. QTCs and Outer Billiards 58
6.5. QTCs and Double Lattice PETs 60

Part 2. Renormalization and Symmetry 63

Chapter 7. Elementary Properties 65
7.1. Intersection of the Parallelograms 66
7.2. Intersection of the Lattices 66
7.3. Rotational Symmetry 66
7.4. Central Tiles 66
7.5. Inversion Symmetry 67
7.6. Insertion Symmetry 67
7.7. The Tiling in Trivial Cases 68

Chapter 8. Orbit Stability and Combinatorics 71
8.1. A Bound on Coefficients 71
8.2. Sharpness 71
8.3. The Arithmetic Graph 72
8.4. Orbit Stability 73
8.5. Uniqueness and Convergence 74
8.6. Ruling out Thin Regions 74
8.7. Joint Convergence 75

Chapter 9. Bilateral Symmetry 77
9.1. Pictures 77
9.2. Definitions and Formulas 78

Chapter 10. Proof of the Main Theorem 81
10.1. Discussion and Overview 81
10.2. Proof of Induction Lemma I 82
10.3. Proof of Induction Lemma II 84

Chapter 11. The Renormalization Map 87
11.1. Elementary Properties 87

11.2. The Even Expanson 87
11.3. Oddly Even Numbers 88
11.4. The Even Expansion and Continued Fractions 89
11.5. Diophantine Approximation 89
11.6. Dense Orbits 90
11.7. Proof of the Triangle Lemma 91

Chapter 12. Properties of the Tiling 93
12.1. Tedious Special Cases 93
12.2. Classification of Tile Shapes 93
12.3. Classification of Stable Orbits 95
12.4. Existence of Square Tiles 96
12.5. The Oddly Even Case 97
12.6. Density of Shapes 97

Part 3. Metric Properties 99

Chapter 13. The Filling Lemma 101
13.1. The Layering Constant 101
13.2. The Filling Lemma, Part 1 101
13.3. The Filling Lemma, Part 2 104

Chapter 14. The Covering Lemma 105
14.1. The Main Result 105
14.2. Some Additional Pictures 110

Chapter 15. Further Geometric Results 111
15.1. The Area Lemma 111
15.2. Tiles in Symmetric Pieces 112
15.3. Pyramids 113

Chapter 16. Properties of the Limit Set 115
16.1. Elementary Topological Properties 115
16.2. Zero Area 116
16.3. Projections of the Limit Set 116
16.4. Finite Unions of Lines 118
16.5. Existence of Aperiodic Points 119
16.6. Hyperbolic Symmetry 119

Chapter 17. Hausdorff Convergence 121
17.1. Results about Patches 121
17.2. Convergence 121
17.3. Covering 123

Chapter 18. Recurrence Relations 127

Chapter 19. Hausdorff Dimension Bounds 131
19.1. The Upper Bound Formula 131
19.2. A Formula in the Oddly Even Case 131
19.3. One Dimensional Examples 132
19.4. A Warm-Up Case 133

19.5. Most of The General Bound 133
19.6. Dealing with the Exceptions 136

Part 4. Topological Properties 139

Chapter 20. Controlling the Limit Set 141
20.1. The Shield Lemma 141
20.2. Another Version of the Shield Lemma 143
20.3. The Pinching Lemma 145
20.4. Rational Oddly Even Parameters 147

Chapter 21. The Arc Case 149
21.1. The Easy Direction 149
21.2. A Criterion for Arcs 150
21.3. Elementary Properties of the Limit Set 153
21.4. Verifying the Arc Criterion 153

Chapter 22. Further Symmetries of the Tiling 157
22.1. Zones 157
22.2. Symmetry of Zones 158
22.3. Intersections with Zones 158
22.4. Folding 160

Chapter 23. The Forest Case 163
23.1. Reduction to the Loops Theorem 163
23.2. Proof of the Loops Theorem 163
23.3. An Example 164

Chapter 24. The Cantor Set Case 167
24.1. Unlikely Sets 167
24.2. Tails and Anchored Paths 167
24.3. Acute Crosscuts 168
24.4. The Main Argument 172
24.5. Pictorial Explanation 173

Chapter 25. Dynamics in the Arc Case 175
25.1. The Main Result 175
25.2. Intersection with the Partitions 176
25.3. The Rational Case 178
25.4. Measures of Symmetric Pieces 179
25.5. Controlling the Measures 180
25.6. The End of the Proof 181

Part 5. Computational Details 183

Chapter 26. Computational Methods 185
26.1. The Fiber Bundle Picture 185
26.2. Avoiding Computational Error 186
26.3. Dealing with Polyhedra 187
26.4. Verifying the Partition 188
26.5. Verifying Outer Billiards Orbits 189

26.6. A Planar Approach 190
26.7. Generating the Partitions 191

Chapter 27. The Calculations 193
27.1. Calculation 1 193
27.2. Calculation 2 194
27.3. Calculation 3 194
27.4. Calculation 4 196
27.5. Calculation 5 196
27.6. Calculation 6 197
27.7. Calculation 7 198
27.8. Calculation 8 198
27.9. Calculation 9 198
27.10. Calculation 10 198
27.11. Calculation 11 199
27.12. Calculation 12 200

Chapter 28. The Raw Data 203
28.1. A Guide to the Files 203
28.2. The Main Domain 203
28.3. The Symmetric Pieces 203
28.4. Period Two Tiles 204
28.5. The Domains from the Main Theorem 204
28.6. The Polyhedra in the Partition 204
28.7. The Action of the Map 206
28.8. The Partition for Calculation 11 207
28.9. The First Partition for Calculation 12 208
28.10. The Second Partition for Calculation 12 209

Bibliography 211

Preface

Polytope exchange transformations are higher dimensional generalizations of interval exchange transformations, one dimensional maps which have been extensively and very fruitfully studied for the past 40 years or so. Polytope exchange transformations have the added appeal that they produce intricate fractal-like tilings. At this point, the higher dimensional versions are not nearly as well understood as their 1-dimensional counterparts, and it seems natural to focus on such questions as finding a robust renormalization theory for a large class of examples.

In this monograph, we introduce a general method of constructing polytope exchange transformations (PETs) in all dimensions. Our construction is functorial in nature. One starts with a multigraph such that the vertices are labeled by convex polytopes and the edges are labeled by Euclidean lattices in such a way that each vertex label is a fundamental domain for all the lattices labeling incident edges. There is a functor from the fundamental groupoid of this multigraph into the category of PETs, and the image of this functor contains many interesting examples. For instance, one can produce huge multi-parameter families based on finite reflection groups.

Most of this monograph is devoted to the study of the simplest examples of our construction. These examples are based on the order 8 dihedral reflection group D_4. The corresponding multigraph is a digon (two vertices connected by two edges) decorated by 2-dimensional parallelograms and lattices. This input produces a 1-parameter family of polygon exchange transformations which we call the Octagonal PETs. One particular parameter is closely related to a system studied by Adler-Kitchens-Tresser.

We show that the family of octagonal PETs has a renormalization scheme in which the $(2, 4, \infty)$ hyperbolic reflection triangle group acts on the parameter space (by linear fractional transformations) as a renormalization symmetry group. The underlying hyperbolic geometry symmetry of the system allows for a complete classification of the shapes of the periodic tiles and also a complete classification of the topology of the limit sets.

We also establish a local equivalence between outer billiards on semi-regular octagons and the octagonal PETs, and this gives a similarly complete description of outer billiards on semi-regular octagons. Finally, we show how the octagonal PETs arise naturally as invariant slices of certain 4-dimensional PETs based on deformations of the E_4 lattice.

I discovered almost all the material in this monograph by computer experimentation, and then later on found rigorous proofs. Most of the proofs here are traditional, but they do rely on 12 computer calculations. These calculations are described in detail in the last part of the monograph.

I wrote two interactive Java programs, OctaPET and BonePET, which illustrate essentially all the mathematics in this monograph. The reader can download these programs from my website (as explained at the end of the introduction) and can use them while reading the manuscript. I wrote the monograph with the intention that a serious reader would also use the programs.

I would like to thank Nicolas Bedaride, Pat Hooper, Injee Jeong, John Smillie, and Sergei Tabachnikov for interesting conversations about topics related to this work. Some of this work was carried out at ICERM in Summer 2012, and most of it was carried out during my sabbatical at Oxford in 2012–2013. I would especially like to thank All Souls College, Oxford, for providing a wonderful research environment.

My sabbatical was funded from many sources. I would like to thank the National Science Foundation, All Souls College, the Oxford Maths Institute, the Simons Foundation, the Leverhulme Trust, the Chancellor's Professorship, and Brown University for their support during this time period.

Oxford, November 2012

CHAPTER 1

Introduction

1.1. What is a PET?

We begin by defining the main objects of study in this monograph. §2 gives more information about what we say here.

PETs: A *polytope exchange transformation* (or PET) is defined by a big polytope X which has been partitioned in two ways into small polytopes:

$$X = \bigcup_{i=1}^{m} A_i = \bigcup_{i=1}^{m} B_i. \tag{1.1}$$

What we mean is that, for each i, the two polytopes A_i and B_i are translation equivalent. That is, there is some vector V_i such that $B_i = A_i + V_i$. We always take the small polytopes to be convex, but sometimes X will not be convex.

We define a map $f : X \to X$ and its inverse $f^{-1} : X \to X$ by the formulas

$$f(x) = x + V_i \quad \forall x \in \text{int}(A_i), \qquad f^{-1}(y) = y - V_i \quad \forall y \in \text{int}(B_i). \tag{1.2}$$

f is not defined on points of ∂A_i and f^{-1} is not defined for points in ∂B_i. Even though f and f^{-1} are not everywhere defined, almost every point of x has a well-defined forwards and backwards orbit.

The Periodic Tiling: A point $p \in X$ is called *periodic* if $f^n(p) = p$ for some n. We will establish the following well-known results in §2: If p is a periodic point, then there is a maximal open convex polytope $U_p \subset X$ such that $p \in U_p$, and $f, ..., f^n$ are entirely defined on U_p, and every point of U_p is periodic with period n. We call U_p a *periodic tile*. We let Δ denote the union of periodic tiles. We call Δ the *periodic tiling*.

The Limit Set: When Δ is dense in X – and this happens in the cases of interest to us here – the *limit set* Λ consists of those points p such that every neighborhood of p intersects infinitely many tiles of Δ. Sometimes Λ is called the *residual set*. See §2.5 for a more general definition.

The Aperiodic Set: We let $\Lambda' \subset \Lambda$ denote the union of points with well-defined orbits. These orbits are necessarily aperiodic, so we call Λ' the *aperiodic set*.

1.2. Some Examples

The simplest examples of PETs are 1-dimensional systems, known as *interval exchange transformations* (IETs). Such a system is easy to produce: Partition an interval smaller intervals, then rearrange them. IETs have been extensively studied in the past 40 years. One very early paper is [**K**]; see papers [**Y**] and [**Z**] for surveys of the literature. The *Rauzy induction* [**R**] gives a satisfying renormalization theory for the family of IETs all having the same number of intervals in the partition. (There are other renormalization schemes as well.)

The simplest examples of higher dimensional polytope exchange transformations are products of IETs. In this case, all the polytopes are rectangular solids. More generally, one can consider PETs (not necessarily products) in which all the polytopes are rectangular solids. In 2 dimensions, these are called *rectangle exchanges*. The paper [**H**] establishes some foundational results about rectangle exchanges.

The paper [**AKT**] gives some early examples of piecewise isometric maps which are not rectangle exchanges. The main example analyzed in [**AKT**] produces locally the same tiling as outer billiards on the regular octagon, and also the same tiling as one of the examples studied here.

The papers [**T2**], [**AE**], [**Go**], [**Low1**], and [**Low2**] all treat a closely related set of systems with 5-fold symmetry which produce tilings by regular pentagons and/or regular decagons. The papers [**AG**], [**LKV**], [**Low1**] and [**Low2**] deal with the case of 7-fold symmetry, which is much more difficult. The difficulty comes from the fact that $\exp(2\pi i/7)$ is a cubic irrational, though one case with 7-fold symmetry is analyzed completely in [**Low2**].

Outer billiards on the regular n-gon furnishes an intriguing family of PETs. The cases $n = 3, 4, 6$ produce regular tilings of the plane, and the cases $n = 5, 7, 8, 10, 12$, where $\exp(2\pi/n)$ is a quadratic irrational, can be completely understood in terms of renormalization. See [**T2**] for the case $n = 5$, and [**BC**] for the other cases. There is partial information about the case $n = 7$, and the remaining cases are not understood at all.

Some definitive theoretical work concerning the entropy of PETs is done in [**GH1**], [**GH2**], and [**B**]. The main results here are that such systems have zero entropy, with a suitable definition of entropy.

The recent paper [**Hoo**] is very close in spirit to our work here. In [**Hoo**], the author works out a renormalization scheme for a 2-parameter family of (non-product) rectangle exchange transformations.

1.3. Goals of the Monograph

Multigraph PETs The first goal of this monograph is to give a general construction of PETs, based on decorated multigraphs. A *multigraph* is a graph in which one allows multiple edges connecting different vertices. The vertices are labelled by convex polytopes and the edges are labeled by Euclidean lattices, so that a vertex is incident to an edge iff the corresponding polytope is a fundamental domain for the corresponding lattice. Given such a multigraph G, we choose a base vertex x and we construct a functorial homomorphism

$$\pi_1(G, x) \to \mathrm{PET}(X). \tag{1.3}$$

Here $\pi_1(G, x)$ is the fundamental group of G, and $\mathrm{PET}(X)$ is the group of PETs whose domain is the polytope X corresponding to x. We call the resulting systems *multigraph PETs*. When G is a digon–i.e. two vertices connected by two edges, we call the system a *double lattice PET*. We will give a variety of nontrivial constructions of multigraph PETs, some related to outer billiards as in [**S2**] and some based on finite reflection groups.

Structure of the Octagonal PETs: The octagonal PETs are the simplest example of our construction. They are planar double lattice PETs based on the order 8 dihedral reflection group. The second, and main, goal of this monograph is study the structure of the octagonal PETs. We state all our main results about the octagonal PETs in the sections following this one.

Connection to Outer Billiards: The third goal is to connect the octagonal PETs of outer billiards on semi-regular octagons. We will prove that outer billiards relative to any semi-regular octagon produces a periodic tiling locally isometric to the one produced by an octagonal PET at a suitable parameter. This fact allows us to give very detailed information about outer billiards on semi-regular octagons.

Connection to the Alternating Grid System: The fourth goal is to explain the connection between the octagonal PETs and a certain dynamical system in the plane (described below) that is defined by a pair of square grids. Each alternating grid system has a 4-dimensional compactification which is a double lattice PET, and the PET has an invariant 2-dimensional slice which is the octagonal PET at the same parameter. This is how we found the octagonal PETs.

1.4. The Octagonal PETs

Here we describe the octagonal PETs. Our construction depends on a parameter $s \in (0, \infty)$, but usually we take $s \in (0,1)$. We suppress s from our notation for most of the discussion.

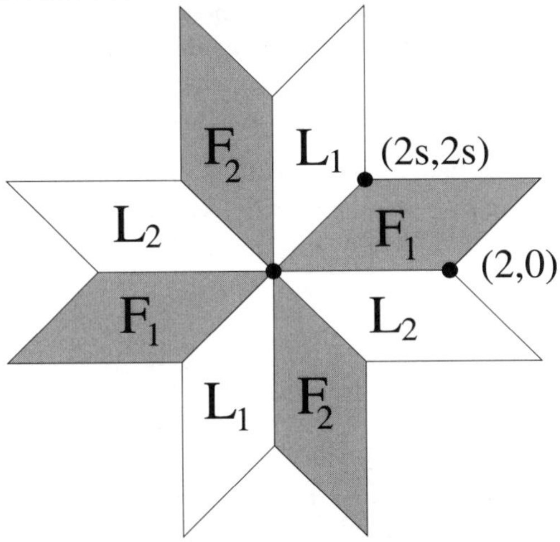

Figure 1.1: The scheme for the PET.

The 8 parallalograms in Figure 1.1 are the orbit of a single parallelogram under a dihedral group of order 8. Two of the sides of this parallelogram are determined by the vectors $(2,0)$ and $(2s, 2s)$. For $j = 1, 2$, let F_j denote the parallelogram centered at the origin and translation equivalent to the ones in the picture labeled F_j. Let L_j denote the lattice generated by the sides of the parallelograms labeled L_j. (Either one generates the same lattice.)

In §2 we will check the easy fact that F_i is a fundamental domain for L_j, for all $i, j \in \{1, 2\}$. We define a system (X', f'), with $X' = F_1 \cup F_2$, and $f' : X' \to X'$, as follows. Given $p \in F_j$ we let

$$f'(p) = p + V_p \in F_{3-j}, \qquad V_p \in L_{3-j}. \tag{1.4}$$

The choice of V_p is almost always unique, on account of F_{3-j} being a fundamental domain for L_{3-j}. When the choice is not unique, we leave f' undefined. When $p \in F_1 \cap F_2$ we have $V_p = 0 \in L_1 \cap L_2$. We will show in §2 that (X', f') is a PET.

We prefer the map $f = (f')^2$, which preserves both F_1 and F_2. We set $X = F_1$. Our system is $f : X \to X$, which we denote by (X, f).

1.5. The Main Theorem: Renormalization

We define the *renormalizaton map* $R : (0, 1) \to [0, 1)$ as follows.

- $R(x) = 1 - x$ if $x > 1/2$
- $R(x) = 1/(2x) - \text{floor}(1/(2x))$ if $x < 1/2$.

R relates to the $(2, 4, \infty)$ reflection triangle triangle much in the way that the classical Gauss map

$$g(x) = 1/x - \text{floor}(1/x)$$

relates to the modular group. We will discuss the hyperbolic geometry connection in §16.6. The map R is an infinite-to-one map of the interval. Almost every orbit of R is dense, and the quadratic irrational numbers are all periodic points of R.

Define

$$Y = F_1 - F_2 = X - F_2 \subset X. \tag{1.5}$$

Figure 1.2 shows a picture of the set Y in one case.

For any subset $S \subset X$, let $f|S$ denote the first return map to S, assuming that this map is defined. When we use this notation, it means implicitly that the map is actually defined, at least away from a finite union of line segments. We call S *clean* if no point on ∂S has a well defined orbit. This means, in particular, that no tile of Δ crosses over ∂S.

THEOREM 1.1 (Main). *Suppose $s \in (0, 1)$ and $t = R(s) \in (0, 1)$. There is a clean set $Z_s \subset X_s$ such that $f_t|Y_t$ is conjugate to $f_s^{-1}|Z_s$ by a map ϕ_s.*

1. *ϕ_s commutes with reflection in the origin and maps the acute vertices of X_t to the acute vertices of X_s.*
2. *When $s < 1/2$, the restriction of ϕ_s to each component of Y_t is an orientation reversing similarity, with scale factor $s\sqrt{2}$.*
3. *When $s < 1/2$, either half of ϕ_s extends to the trivial tile of Δ_t and maps it to a tile in Δ_s which has period 2.*
4. *When $s < 1/2$, the only nontrivial orbits which miss Z_s are contained in the ϕ_s-images of the trivial tile of Δ_t. These orbits have period 2.*
5. *When $s > 1/2$ the restriction of ϕ_s to each component of Y_s is a translation.*
6. *When $s > 1/2$, all nontrivial orbits intersect Z_s.*

Our main result is an example of a situation where a picture says a thousand words. Figures 1.2 and 1.3 show the Main Theorem in action for $s < 1/2$. Figures 1.4 and 1.5 show the Main Theorem in action for $s > 1/2$.

Figure 1.2: Y_t lightly shaded for $t = 3/10 = R(5/13)$.

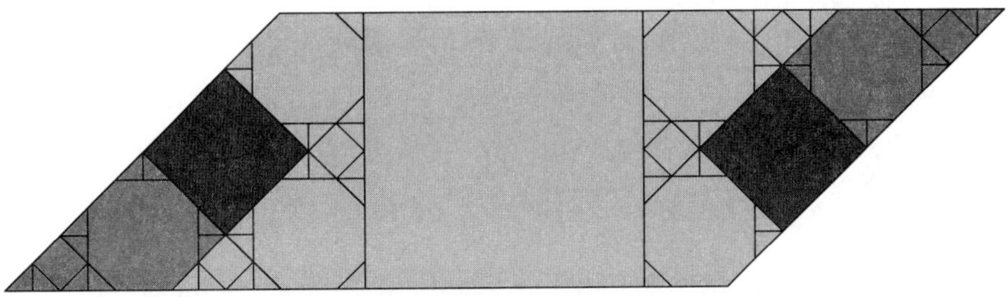

Figure 1.3: Z_s lightly shaded $s = 5/13$.

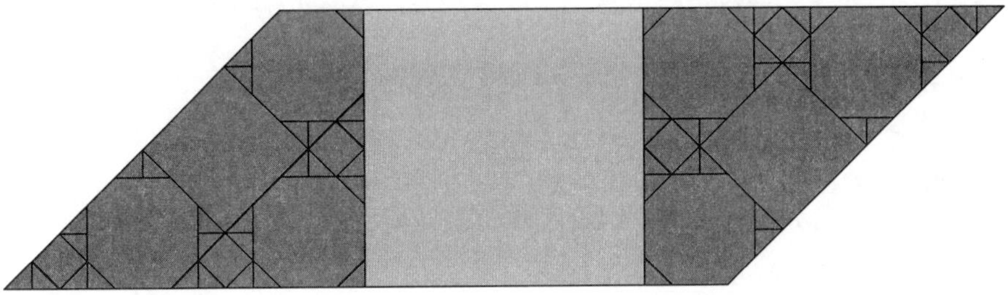

Figure 1.4: Y_t lightly shaded for $t = R(8/13) = 5/13$.

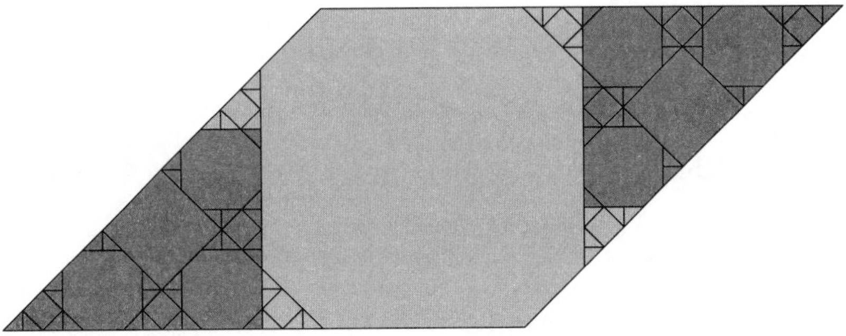

Figure 1.5: Z_s lightly shaded for $s = 8/13$.

1.6. Corollaries of The Main Theorem

1.6.1. Structure of the Tiling. A *semi-regular octagon* is an octagon with 8-fold dihedral symmetry. When $s \in (1/2, 1)$, the intersection

$$O_s = (F_1)_s \cap (F_2)_s \tag{1.6}$$

is the semi-regular octagon with vertices

$$(\pm s, \pm(1-s)), \qquad (\pm(1-s), \pm s). \tag{1.7}$$

When $s \in (0, 1/2)$, the interection O_s is the square with vertices $(\pm s, \pm s)$.

Given $s \in (0,1)$, let $s_n = R^n(s)$. We call the index n *good* if $s_{n-1} < 1/2$ or if $n = 0$.

THEOREM 1.2. *When s is irrational, a polygon appears in Δ_s if and only it is similar to O_{s_n} for a good index n. When s is rational, a polygon appears in Δ_s if and only if it is a square, a right-angled isosceles triangle, or similar to O_{s_n} for a good index n.*

Remark: Theorem 1.2 is a consequence of a more precise and technical result, Theorem 12.1, which describes the tiles of Δ_s up to translation.

According to Theorem 1.2, the tiling Δ_s is an infinite union of squares and semi-regular octagons when s is irrational. Here is some more information in the irrational case.

THEOREM 1.3. *The following is true when $s \in (0,1)$ is irrational.*

1. *If Δ_s has no squares then $s = \sqrt{2}/2$. If Δ_s has finitely many squares, then $s \in \boldsymbol{Q}[\sqrt{2}]$.*
2. *Δ_s has only squares if and only if the continued fraction expansion of s has the form $(0, a_0, a_1, a_2, ...)$ where a_k is even for all odd k. This happens if and only if $R^n(s) < 1/2$ for all n.*
3. *Δ_s has infinitely many squares and a dense set of shapes of semi-regular octagons for almost all s.*

According to Theorem 1.2, when s is rational, Δ_s consists entirely of squares, semi-regular octagons, and right-angled isosceles triangles. We say that a periodic tile τ_s of Δ_s is *stable* if, for all parameters r sufficiently close to s, there is a tile τ_r of Δ_r consisting of points having the same period and dynamical behavior as the points in τ_s. See §8 for more information.

THEOREM 1.4. *When s is rational, a tile of Δ_s is unstable if and only if it is a triangle. All the unstable tiles are isometric to each other, and they are arranged into 4 orbits, all having the same period.*

1.6.2. Structure of the Limit Set. Here are some results about the limit set. The quantity $\dim(\Lambda_s)$ denotes the Hausdorff dimension of the limit set.

THEOREM 1.5. *Suppose s is irrational.*

1. *Λ_s has zero area.*
2. *The projection of Λ_s onto a line parallel to any 8th root of unity contains a line segment. Hence $\dim(\Lambda_s) \geq 1$.*
3. *Λ_s is not contained in a finite union of lines.*
4. *Λ'_s is dense in Λ_s.*

Theorem 1.4 implies, if $s \in \boldsymbol{Q}$, that there is a single number $N(s)$ such that Δ_s has $4N(s)$ triangular tiles, and all unstable periodic orbits of (X_s, f_s) have period $N(s)$. In §18, we will give a kind of formula for $N(s)$. We will then use this formula to get some upper bounds on the $\dim(\Lambda_s)$.

THEOREM 1.6. *For any irrational s, there is a sequence $\{p_n/q_n\}$ of rational parameters, converging to s, such that*

$$\dim(\Lambda_s) \leq \limsup \frac{\log N(p_n/q_n)}{\log q_n}. \tag{1.8}$$

In particular

- $\dim(\Lambda_s) = 1$ *if* $\lim R^n(s) = 0$.
- $\dim(\Lambda_s) \leq 1 + (\log 8/\log 9)$ *in all cases.*

Remarks:
(i) The sequence $\{p_n/q_n\}$ in the theorem is closely related to the sequence of continued fraction approximations to s. Technically, we have $p_n/q_n \to s$ in the sense of §11.5.
(ii) Computer experiments with our formula for $N(s)$ suggest that

$$\dim(\Lambda_s) \leq \frac{2\log(1+\sqrt{2})}{\log(2+\sqrt{3})}, \qquad \forall\ s \in (0,1). \tag{1.9}$$

The bound is attained when $s = \sqrt{3}/2 - 1/2$. Figure 1.6 below shows the picture for this parameter. The case when $R^n(s) > 1/2$ finitely often boils down to a question about the dynamics of the Gauss map. In a private communication, Pat Hooper sketched a proof for me that the result holds in this case.
(iii) Theorem 1.6 says, in particular, that $\dim(\Lambda_s) < 2$, which of course implies that Λ_s has zero area. However, we prove that Λ_s has zero area separately because the proof of Theorem 1.6 is rather involved.

Now we turn to questions about the topololy of Λ_s. Here is the complete classification of the topological types.

THEOREM 1.7. *Let $s \in (0,1)$ be irrational.*

1. *Λ_s is a disjoint union of two arcs if and only if Δ_s contains only squares. This happens if and only if $R^n(s) < 1/2$ for all n.*
2. *Λ_s is a finite forest if and only if Δ_s contains finitely many octagons. This happens if and only if $R^n(s) > 1/2$ for finitely many n.*
3. *Λ_s is a Cantor set if and only if Δ_s contains infinitely many octagons. This happens if and only if $R^n(s) > 1/2$ for infinitely many n.*

Figures 1.6-1.8 illustrate the three cases of Theorem 1.7. The curve in Figure 1.6 is isometric to one of the curves which appears in Pat Hooper's Truchet tile systems. The notation in Figure 1.7 refers to the even expansion of s. See §11 for a definition. The tiling in Figure 1.8 is the same as the main example in [**AKT**].

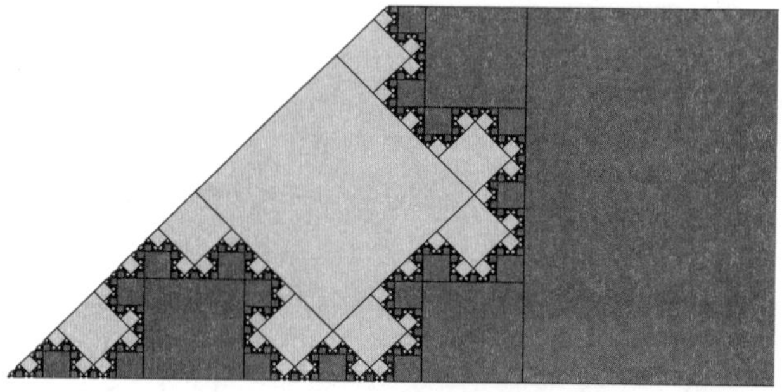

Figure 1.6: The left half of Δ_s for $s = \sqrt{3}/2 - 1/2$.

Figure 1.7: The left half of Δ_s for $s = (0, 3, 1, 3, 1, 2, 2, 2...)$.

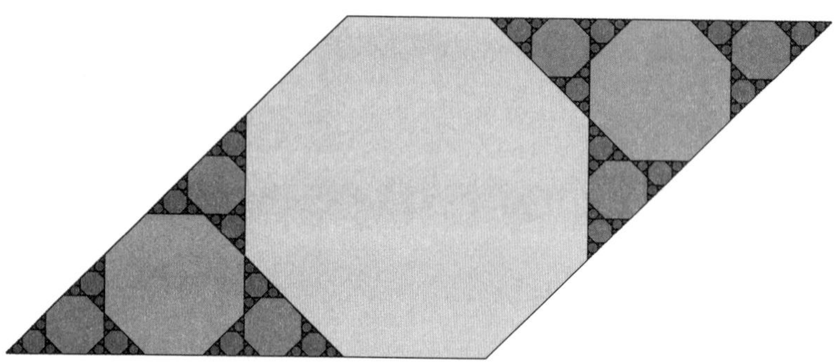

Figure 1.8: The left half of Δ_s for $s = \sqrt{2}/2$.

We will investigate Case 1 of Theorem 1.7 in more detail.

THEOREM 1.8. *Suppose that the continued fraction of s is* $(0, a_1, a_2, a_3, ...)$ *with* a_k *even for all odd k. Then the restriction* $f_s|\Lambda_s$ *is a* $\boldsymbol{Z}/2$ *extension of the irrational rotation with rotation number* $(0, 2a_2, a_3, 2a_4, a_5, 2a_6, a_7, ...)$.

Remark: Theorem 1.8, and the more precise version, Theorem 25.1, are similar to forthcoming results of Pat Hooper. Indeed, conversations with Hooper inspired me to formulate and prove Theorems 1.8 and 25.1.

Now we discuss the dependence of the limit set on the parameter. When s is rational, we let U_s denote the closure of the union of the unstable tiles. Let $\mathcal{K}$ denote the set of compact subsets of $\boldsymbol{R}^2$. We equip $\mathcal{K}$ with the *Hausdorff metric*.

The distance between two subsets is the infimal ϵ such that each is contained in the ϵ-tubular neighborhood of the other.

We have a map $\Xi : (0,1) \to \mathcal{K}$ defined as follows.

- $\Xi_s = U_s$ when s is rational
- $\Xi_s = \Lambda_s$ when s is irrational.

THEOREM 1.9. *The map Ξ is continuous at irrational points of $(0,1)$.*

Remark: Theorem 1.9 says, in particular, that the union of unstable orbits in the rational case gives a good approximation to the limit set in the irrational case. Theorem 17.9, which is the main tool we use to prove Theorem 1.6, gives another view of this same idea.

1.6.3. Hyperbolic Symmetry.

The Main Theorem above implies that there is an underlying hyperbolic symmetry to the family of octagonal PETs. (See §2.6 for some background information on hyperbolic geometry.)

Let $\boldsymbol{H}^2 \subset \boldsymbol{C}$ denote the upper half plane model of the hyperbolic plane. Let Γ denote the $(2,4,\infty)$ reflection triangle group, generated by reflections in the sides of the ideal hyperbolic triangle with vertices

$$\frac{i}{\sqrt{2}}, \qquad \frac{1}{2}+\frac{i}{2}, \qquad \infty. \tag{1.10}$$

We extend our parameter range so that our system is defined for all $s \in \boldsymbol{R}$. The systems at s and $-s$ are identical. Γ acts on the parameter set by linear fractional transformations.

We say that (X_s, f_s) is *locally modelled on* (X_t, f_t) at $p_s \in X_s$ if there is some $p_t \in X_t$ and a similarity $g : \Delta_s \cap U_s \to \Delta_t \cap U_t$, with U_s and U_t being neighborhoods of p_s and p_t respectively. We say that (X_s, f_s) and (X_t, f_t) are *locally equivalent* if there are finite collections of lines L_s and L_t such that (X_s, f_s) is locally modelled on (X_t, f_t) for all $p_t \in \Lambda_t - L_t$ and (X_t, f_t) is locally modelled on (X_t, f_s) for all $p_s \in \Lambda_s - L_s$. Intuitively, the tilings of locally equivalent systems have the same fine-scale structure. In particular, the limit sets of locally equivalent systems have the same Hausdorff dimension.

THEOREM 1.10. *Suppose s and t are in the same orbit of Γ. Then (X_s, f_s) and (X_t, f_t) are locally equivalent. In particular, the function $s \to \dim(\Lambda_s)$ is a Γ-invariant function.*

Remarks:
(i) Γ is contained with index 4 in the group generated by reflections in the ideal triangle with vertices $0, 1, \infty$. Using this fact, together with a classic result about continued fractions, we we will show that the forward orbit $\{R^n(s)\}$ is dense in $(0,1)$ for almost all $s \in (0,1)$.
(ii) The need to exempt a finite union of lines in the definition of local equivalence seems partly to be an artifact of our proof, but in general one needs to disregard some points to make everything work.
(iii) Given the ergodic nature of the action of Γ, we can say that there is some number δ_0 such that $\dim(\Lambda_s) = \delta_0$ for almost all s. However, we don't know the value of δ_0.

1.7. Polygonal Outer Billiards

B. H. Neumann [**N**] introduced outer billiards in the late 1950s and J. Moser [**M1**] popularized the system in the 1970s as a toy model for celestial mechanics. Outer billiards is a discrete self-map of $\boldsymbol{R}^2 - P$, where P is a bounded convex planar set as in Figure 1.9 below. Given $p_1 \in \boldsymbol{R}^2 - P$, one defines p_2 so that the segment $\overline{p_1p_2}$ is tangent to P at its midpoint and P lies to the right of the ray $\overrightarrow{p_1p_2}$. The map $p_1 \to p_2$ is called *the outer billiards map.*

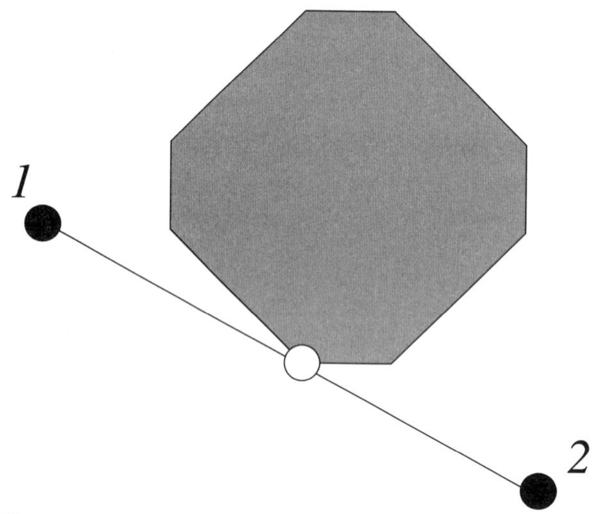

Figure 1.9: Outer billiards relative to a semi-regular octagon.

The second iterate of the outer billiards map is a PET whose domain is the entire plane. (We make the map the identity inside the central shape.) In particular, we define $\Delta(P)$ and $\Lambda(P)$ as above.

Much work on outer billiards has been done in recent years. See, for instance, [**BC**], [**Bo**], [**DT1**], [**DT2**], [**G**], [**GS**], [**Ko**], [**S2**], [**S3**], [**T1**], [**T2**], and [**VS**]. There is a survey in [**T1**], and a more recent one in [**S3**].

The sets $\Delta(P)$ and $\Lambda(P)$ are well understood when P is a regular n-gon for $n = 3, 4, 5, 8, 12$. These cases correspond to some of the piecewise isometric systems discussed above. (The case $n = 12$ has not really been worked out, but it is quite similar to the cases $n = 5, 8$.) Many pictures have been drawn for other values of n, but there are no theoretical results.

I have given some information in [**S4**] about $\Delta(P)$ when P is the Penrose kite, but the information is far from complete. In all other cases, nothing is known about the tiling and the limit set.

The only thing known about outer billiards on semi-regular octagons is that all orbits are bounded. This follows from the general result in [**GS**], [**Ko**], and [**VS**]. Here we will give fairly complete information about outer billiards on semi-regular octagons.

We parameterize semi-regular octagons as in Equation 1.6. The complement $\boldsymbol{R}^2 - O_s$ is tiled, in two ways, by isometric copies of $X_s = (F_1)_s$. The two ways are mirror reflections of each other. We first divide the $\boldsymbol{R}^2 - O_s$ into 8 cones, and then we fill each cone with parallel copies of X_s, in the pattern (partially) shown.

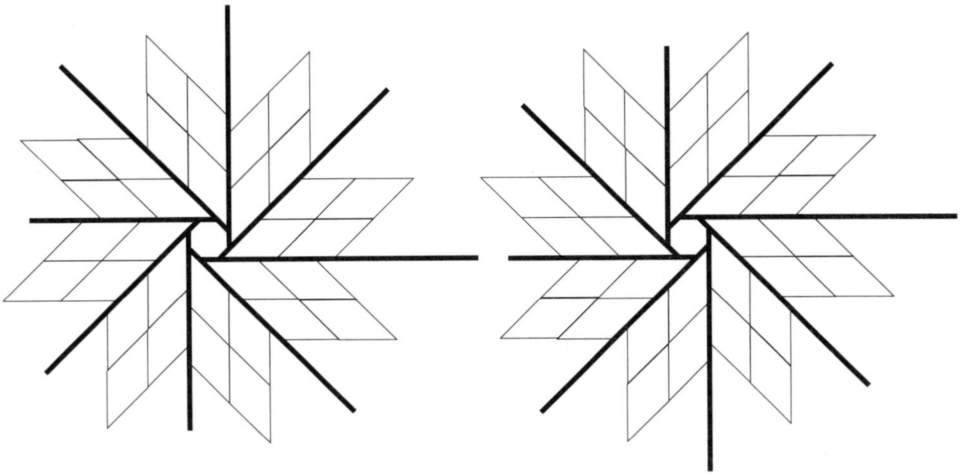

Figure 1.10: The tiling of $\boldsymbol{R}^2 - O_s$.

THEOREM 1.11. *Suppose $s \in (1/2, 1)$. Let Y be any parallelogram in either of the two tilings of $\boldsymbol{R}^2 - O_s$. Let Δ_s and Λ_s be the periodic tiling and the limit set for the octagonal PET (X_s, f_s). Then $\Delta(O_s) \cap Y$ and $\Lambda(O_s) \cap Y$ are isometric copies of Δ_s and Λ_s respectively.*

Thanks to Theorem 1.11, all the results mentioned above for the octagonal PETs have analogous statements for outer billiards on semi-regular octagons. For instance, almost every point is periodic and the Hausdorff dimension of the limit set is strictly less than 2. The map R does not preserve the interval $(1/2, 1)$ and indeed there are many points in $(1/2, 1)$, such as $s = 3/2 - \sqrt{3}/2$, for which $R^n(s)$ never returns to $(1/2, 1)$. Thus, the family of outer billiards systems in semi-regular octagons really only has a renormalization scheme associated to it when it is included in the larger family of octagonal PETs.

In §3.4 we will see that the octagonal PETs correspond to the case $n = 4$ of what we call *dihedral PETs*, a more general construction which works for $n = 3, 4, 5, \ldots$. The case $n = 3$, corresponding to outer billiards on semi-regular hexagons, should have a theory very similar to what we do for $n = 4$ in this monograph.

1.8. The Alternating Grid System

Let $G_{s,z}$ denote the grid of squares in the plane, with side length s, having $z \in \boldsymbol{C}$ as a vertex. We take the sides of the squares to be parallel to the coordinate axes. Let $T_{s,z}$ denote the piecewise isometric transformation which rotates each square of $G_{s,z}$ counterclockwise by $\pi/2$ about its center. The map $T_{s,z}$ is not defined on the edges of $G_{s,z}$.

The alternating grid system is based on the composition of two maps like this, based on differently sized grids. The standard grid is $G_{1,0}$, and we consider the infinite group $\langle T_{1,0}, T_{s,z} \rangle$. This group acts on the plane by piecewise isometries. To get a dynamical system in the traditional sense, one can choose a word

$$T_{1,0}^{e_1} T_{s,z}^{e_2} \cdots T_{1,0}^{e_{n-1}} T_{s,z}^{e_n}, \qquad e_1, \ldots, e_n \in \{0, 1, 2, 3\} \tag{1.11}$$

and study the action of this map on $\boldsymbol{R}^2$. It seems reasonable to require that the exponents sum to 0 mod 4. With this restriction, the map is a PET whose domain is the whole plane.

It is interesting to vary the choice of word and see how the system changes. In [**S6**] we consider this question in detail, and show that most words lead to systems having very few periodic points. We make a case in [**S6**] that the word

$$F_{s,z} = T_{1,0}T_{s,z}T_{1,0}T_{s,z} \tag{1.12}$$

is the most interesting word to study. We call the system $(\boldsymbol{R}^2, F_{s,z})$ the *Alternating Grid System*, or *AGS* for short. When we discuss the AGS in §4, we will present some interesting experimental observations about these systems which go beyond what we can actually prove.

In [**S6**] (which I wrote after completing the first version of this monograph) we show that the noncompact system given by the map in Equation 1.11 always has a higher dimensional compactification, in the sense discussed below. In the case of the AGS, the compactification is 4 dimensional. This is the case we discuss here. The compactification of the system $(\boldsymbol{R}^2, F_{s,z})$ does not depend on z, though the map into the compactification does depend (mildly) on z.

THEOREM 1.12. *The system $(\boldsymbol{R}^2, F_{s,z})$ has a 4-dimensional compactification $(\widetilde{X}_s, \widetilde{F}_s)$, where $\widetilde{X}_s \subset \boldsymbol{R}^4$ is a parallelotope and $\widetilde{F}_s$ is a polytope exchange transformation.*

By *compactification*, we mean that there is an injective piecewise affine map $\widetilde{\Psi}_{s,z} : \boldsymbol{R}^2 \to \widetilde{X}_s$ such that

$$\widetilde{F}_s \circ \Psi_{s,z} = \widetilde{\Psi}_{s,z} \circ F_{s,z} \tag{1.13}$$

When s is irrational, $\Psi_{s,z}$ has a dense image. This map depends on z, but any two choices of z lead to maps which differ by a translation of $\widetilde{X}_s$. We think of $\boldsymbol{R}^2$ as an "irrational plane" sitting inside $\widetilde{X}_s$, as fundamental domain for a 4-dimensional torus. (The choice of plane depends on z.) The map $\widetilde{F}_{s,z}$ acts in such a way as to preserve this irrational plane, and thus induces the action of $F_{s,z}$ on $\boldsymbol{R}^2$.

The pair $(\widetilde{X}_s, \widetilde{F}_s)$ is a 4-dimensional double lattice PET. The next result says that the octagonal PET (X_s, f_s) is the invariant slice of the 4-dimensional compactification of the AGS.

THEOREM 1.13. *The system $(\widetilde{X}_s, \widetilde{F}_s)$ commutes with an involution I_s of $\widetilde{F}_s$. The two eigen-planes of I_s are invariant under the system, and the restriction of $\widetilde{F}_s$ to each one is a copy of the octagonal PET at parameter s.*

Theorem 1.13 combines with Theorem 1.5 to prove the following result.

THEOREM 1.14. *For every irrational $s \in (0,1)$, there is some choice of z (depending on s) such that $(\boldsymbol{R}^2, F_{s,z})$ has unbounded orbits.*

Let $K(s)$ denote the set of z such that $(R^2, F_{s,z})$ has unbounded orbits. Our result above says that $K(s)$ is nonempty. We think that $K(s) = \boldsymbol{C}$.

Combining Theorem 1.13 and Theorem 1.11, we get the following result.

COROLLARY 1.15. For any $s \in (1/2, 1)$ the tiling produced by outer billiards on the semi-regular octagon O_s is locally isometric to the one which appears in the invariant slice of the system $(\widetilde{X}_s, \widetilde{F}_s)$.

In short, one "sees" outer billiards on semi-regular octagons inside the AGS.

So far, I do not have a good understanding of the whole AGS, but I think that the AGS is quite rich and mysterious. It certainly deserves further study. This monograph is really an outgrowth of my attempt to understand the AGS. The few results here about the AGS should really just be the beginning of the story.

1.9. Computer Assists

The proofs we give in Parts I-IV of the monograph rely on 12 computer calculations, which we explain in Part V. These calculations involve showing that various pairs of convex integer polyhedra are either disjoint or nested. Everything is done with integer arithmetic, so that there is no roundoff error. We control the sizes of the integers, so that there is also no overflow error.

My interactive java programs OctaPET and BonePET [1] do all the calculations. OctaPET does all the calculations connected to our main result, and BonePET does all the calculations connected to Theorem 1.11. The programs can be downloaded from the URLs
http://www.math.brown.edu/~res/Java/OCTAPET.tar
http://www.math.brown.edu/~res/Java/BONEPET.tar
These are tarred directories, which untar to a directories called OctaPET and BonePET. These directories contain the programs, as well as instructions for compiling and running them.

All the pictures shown in this monograph are taken from these programs. We strongly advise the reader to use OctaPET and BonePET while reading the monograph. In the monograph we can only illustrate the important phenomena with a few pictures whereas the reader can see the picture for essentially any parameter using the programs. For the sake of giving a readable exposition, we omit a number of routine geometric calculations. Such calculations are all exercises in plane geometry. Rather than write out these calculations, we illustrate them with pictures from our programs. Again, such calculations will be all the more obvious to the reader who is using the programs while looking at the monograph.

Finally, we mention that our proof of Theorem 1.6 requires a small amount of Mathematica [**W**] code, which we include in the a directory called **Mathematica** in the source code for our program OctaPET.

1.10. Organization

Following a chapter containing some background material, this monograph is organized into 5 parts.

- Part I deals with the relation of the octagonal PET to other PETs. In particular, we prove Theorems 1.11, 1.12, and 1.13. We also introduce the multigraph PETs, which are a functorial way of producing PETs.
- In Part II we establish some elementary properties of the octagonsl PETs and then prove the Main Theorem. Following a discussion of the properties of the renormalization map R from the Main Theorem, we prove Theorems 1.2, 1.3, and 1.4.
- In Part III we elaborate on the Main Theorem, by explaining more precisely how the tiling at the parameter s is related to the parameter $R(s)$. We use these results to prove Theorems 1.5, 1.6, 1.9, and 1.10,

[1] My daughter helped me name the second program. The name derives from the fact that the domains for the PETs we use for Theorem 1.11 look like dog bones.

- In Part IV we investigate the topology of the limit set, and at the end, some of the dynamics on the limit set. In particular, we prove the Theorems 1.7 and 1.8,
- In Part V, we present all the computer assisted calculations.

At the beginning of each part of the monograph, we will give a more detailed overview of that part.

Part 1

Friends of the Octagonal PETs

Here is an overview of this part of the monograph:

- In §3 we introduce the multigraph PETs. The typical input to our construction is a finite group of isometries of $\boldsymbol{R}^n$ generated by involutions, and a parallelotope that is adapted to the generators. The octagonal PETs are the simplest nontrivial 2-dimensional examples; they are based on the order 8 dihedral group acting on the plane.
- In §4 we introduce the alternating grid system and prove Theorems 1.12 and 1.13. At the end of §4 we recognize the 4-dimensional compactification of any alternating grid system as one of the PETs constructed in §3.
- In §5 we prove Theorem 1.11, which relates the octagonal PETs to outer billiards on semiregular octagons. Our proof relies on 4 calculations, Calculations 9-12, which we perform in Part IV of the monograph. (Calculations 1-8 deal with properties of the octagonal PET. To some extent, Calculations 9-12 rely on some of them.)
- In §6, a chapter which mainly reports on work done in [**S2**] and [**S4**], we introduce a system called a quarter turn composition (QTC). Such a system is, in some sense, a generalization of the alternating grid system. We show how quarter turn systems are related to decorated multigraphs, to outer billiards, and to double lattice PETs. Tracing through the chain of relations, we see how polygonal outer billiards is related to double lattice PETs.

CHAPTER 2

Background

2.1. Lattices and Fundamental Domains

Here are some basic geometric objects we consider in this monograph.

Convex Polytopes: Just for the record, we say that a *convex polytope* is the convex hull of (i.e. the smallest closed convex set which contains) a finite union of points in Euclidean space. The intersection of finitely many convex polytopes is again a convex polytope, possibly of lower dimension.

We will not bother to define a general (non-convex) polytope, because all the polytopes we consider are convex, except for several explicit 2-dimensional examples.

Parallelotopes: As a special case, a *parallelotope* in $\boldsymbol{R}^n$ is a convex polytope of the form $T(Q)$, where Q is an n-dimensional cube in $\boldsymbol{R}^n$ and T is an invertible affine transformation of $\boldsymbol{R}^n$.

Euclidean Lattices: We define a *lattice* in $\boldsymbol{R}^n$ to be a discrete abelian group of the form $T(\boldsymbol{Z}^n)$, where T is an invertible linear transformtion. When L is a lattice in $\boldsymbol{R}^n$, the quotient $\boldsymbol{R}^n/L$ is a flat torus.

Fundamental Domains: Let L be a lattice. We say that a convex polytope F is a *fundamental domain* for L if the union

$$\bigcup_{V\in L}(F+V) \tag{2.1}$$

is a tiling of $\boldsymbol{R}^n$. By this we mean that the translates of the interior of F by vectors in L are pairwise disjoint and the translates of F itself cover $\boldsymbol{R}^n$.

Equivalently, we can say that F is a fundamental domain if the following is true.

- For all points $p \in \boldsymbol{R}^n$ there is some vector $V_p \in L$ such that $p + V_p \in F$.
- F and $\boldsymbol{R}^n/L$ have the same volume.

These conditions in turn imply that the choice of V_p is unique unless p lies in a certain countable union of codimension 1 sets – namely the L-translates of the faces of F.

2.2. Hyperplanes

Basic Definition: A *hyperplane* in $\boldsymbol{R}^n$ is a solution V to an equation

$$v \cdot n = d \qquad \forall\, v \in V. \tag{2.2}$$

for some $d \in \boldsymbol{R}$ and some nonzero vector n. The vector n is called a *normal* to V. One might take n to be a unit vector, but this is not necessary. When $d = 0$, the hyperplane V is a vector subspace of $\boldsymbol{R}^n$.

Matrix Action: Suppose that $M : \boldsymbol{R}^n \to \boldsymbol{R}^n$ is an invertible linear transformation. We shall have occasion to want to know how M acts on hyperplanes. Suppose V is the hyperplane satisfying Equation 2.2. Then $M(V)$ is the hyperplane satisfying the equation

$$w \cdot (M^{-1})^t(n) = d \qquad \forall\, v \in M(V) \tag{2.3}$$

This equation derives from the fact that, in general $M(v) \cdot w = v \cdot M^t(w)$.

Parallel Families: We say that a *parallel family* of hyperplanes is a countable discrete set of evenly spaced parallel hyperplanes. The *direction* of the parallel family is specified by giving a normal vector to any hyperplane in the family. All the hyperplanes in the family have the same normal.

Full Families: We say that a collection of n parallel families of hyperplanes is *full* if the corresponding normals form a basis for for $\boldsymbol{R}^n$. The standard example is the collection $C_1, ..., C_n$, where C_i consists of those points $(x_1, ..., x_n)$ with $x_i \in \boldsymbol{Z}$.

LEMMA 2.1. *An arbitrary full family is equivalent to the standard example by some affine map. In particular, the complement of a full family is a periodic tiling by parallelotopes.*

Proof: Let $v_1, ..., v_n$ be a basis of normals, where v_i is normal to the ith family. Let M be a matrix such that $(M^{-1})^t$ carries our basis to the standard basis. Then M maps the ith full family to a family parallel to C_i defined above. Further composing with a diagonal matrix and then translating, we get exactly the map we seek. □

2.3. The PET Category

We define a category **PET** whose objects are convex polytopes and whose morphisms are (equivalence classes) of PETs between them.

PETs: A PET from X to Y is a map of the form given in Equation 1.2, when we have partitions

$$X = \bigcup_{i=1}^{m} A_i, \qquad Y = \bigcup_{i=1}^{m} B_i. \tag{2.4}$$

with A_i and B_i being translates.

Composition: If we have PET maps $f : X \to Y$ and $f' : Y \to Z$ we can compose them and get a PET map $f' \circ f : X \to Z$. Concretely, suppose that $f : X \to Y$ is defined in terms of partitions $\{A_i\}$ and $\{B_i\}$ and $f' : X \to X$ is defined in terms of partitions $\{A'_i\}$ and $\{B'_i\}$. Then $f' \circ f$ is defined in terms of the partitions

$$\{f^{-1}(B'_i) \cap A_j\}, \qquad \{f'(B_i \cap A'_j)\}. \tag{2.5}$$

and equals the obvious composition on each piece.

Equivalence: We call $f_1, f_2 : X \to Y$ *equivalent* if the two maps agree on a convex polytope partition which is a common refinement of the partitions defining the maps. It is easy to see that the relation on PETs from X to Y really is an equivalence relation.

The Category: PET is the category whose objects are convex polytopes and whose morphisms are equivalence classes of PETs between the polytopes. It is easy that the composition defined above respects the equivalence classes. Hence **PET** really is a category.

Remark: Sometimes it is annoying to distinguish between a PET and an equivalence class of PETs. We make the distinction here so that we can interpret equation 1.3 precisely. In practice, however, we will have a concrete PET defined in terms of some explicit partitions.

2.4. Periodic Tiles for PETs

Let (X, f) be a PET, as in §1.1. The map $f : X \to X$ is based on the two partitions of X, as in Equations 1.1 and 1.2.

Let $\mathcal{A}_1$ denote the collection $\{A_1, ..., A_n\}$ of polytopes in the first partition of X and let $\mathcal{B}_1$ denote the collection $\{B_1, ..., B_n\}$ of polytopes in the second partition.

For positive integers $n \geq 2$, we inductively define $\mathcal{A}_n$ to be the collection of polyhedra

$$f^{-1}(f(P) \cap A), \qquad P \in \mathcal{A}_{n-1}, \qquad A \in \mathcal{A}_1. \tag{2.6}$$

The partition $\mathcal{A}_n$ refines the partition $\mathcal{A}_{n-1}$. All these partitions consist of convex polytopes, and the power f^n is defined on the complement of the union

$$\bigcup_{P \in \mathcal{A}_n} \partial P. \tag{2.7}$$

LEMMA 2.2. *There is a codimension* 1 *subset $S \subset X$ such that every point of $X - S$ has a well-defined orbit.*

Proof: Take S' to be the union of all the sets in Equation 2.7. This set has codimension 1, and every point in $X - S'$ has a well-defined forward orbit. There is an analogous set $S'' \subset X$ such that every point in $X - S''$ has a well-defined backwards orbit. We let $S = S' \cup S''$ and we are done. □

LEMMA 2.3. *Let P be an open polytope of $\mathcal{A}_n$. Suppose that some point of P is periodic, with period n. Then all points of P are periodic, with period n.*

Proof: The first n iterates of f "do the same thing" on the interior of each polytope of $\mathcal{A}_n$. More precisely, if p_1 and p_2 lie in the same open polytope in $\mathcal{A}_n$ then

$$f^k(p_1) - f_k(p_2) = p_1 - p_2, \qquad k = 1, ..., n. \tag{2.8}$$

Our lemma follows immediately from this observation. □

COROLLARY 2.4. Suppose that $p \in X$ is a periodic point. Then there exists a maximal open convex polytope U_p such that f is entirely defined and periodic on U_p.

Proof: Let n be the minimal period of p. Since f^n is defined on p, there is some unique convex open polytope U_p of $\mathcal{A}_n$ which contains p in its interior. By the previous result, all points of U_p have period n. By definition, f^n is not defined on ∂U_p, so U_p is the maximal domain having the advertised properties. □

Remark: There is one subtle point about our previous argument which deserves to be made more clear. Suppose, for instance, that there is some n such that every point on which f is defined has period n. It is tempting to then say that f^n is defined, and the identity, on all of X. However, in order for f^n to be defined, all the iterates f^k must be defined for $k = 1, ..., n$. So, we will not say that f^n is defined, and the identity, on all of X. Rather, we will keep to the original convention and say that f^n is defined, and the identity, on the complement of the set in Equation 2.7.

As mentioned in the introduction, the periodic tiling Δ is the union of the open periodic tiles of f. From the results above, each tile of Δ is an open convex polytope belonging to some $\mathcal{A}_n$. We might have made all of the above definitions and arguments in terms of the inverse map f^{-1} and the analogous partitions $\mathcal{B}_n$. Thus, we can say at the same time that each tile of Δ is an open convex polytope belonging to some $\mathcal{B}_n$. We find it more convenient to work with the forward iterates of f, however.

LEMMA 2.5. *If $\{P_k\}$ is any sequence of tiles in Δ, the period of points in P_k tends to ∞ with k.*

Proof: If this is false, then there is a single n such that P_k is an open polytope of $\mathcal{A}_n$ for all k. But $\mathcal{A}_n$ has only finitely many polytopes. □

Remark: It is worth pointing out that Δ might be empty. If one is willing to disregard a countable set of points, one can view an irrational rotation of the circle as a 2-interval IET. Such an IET has no periodic points at all. D. Genin noticed a similar phenomenon for outer billiards relative to irrational trapezoids [**G**]. We are interested in the opposite case, when Δ is dense.

2.5. The Limit Set

In the introduction we gave a definition of the limit set which works for the systems we study in this monograph. Here we give a more robust definition. We call a point $p \in X$ *weakly aperiodic* if there is a sequence $\{q_n\}$ converging to p with the following property. The first n iterates of f are defined on q_n and the points $f^k(q_n)$ for $k = 1, ..., n$ are distinct. We let Λ denote the union of weakly aperiodic points. We call Λ the *limit set*. Some authors call Λ the *residual set*.

Not all points of Λ need to have well-defined orbits. In fact, the set $\Lambda' \subset \Lambda$ of aperiodic points is precisely the set of points of Λ with well-defined orbits. Now we reconcile the definition here with what we said in the introduction.

LEMMA 2.6. *Suppose that the periodic tiling Δ is dense. Then a point belongs to Λ if and only if every open neighborhood of the point contains infinitely many tiles of Δ.*

Proof: Let Λ^* denote the set of points p such that every neighborhood of p contains infinitely many tiles of Δ. We want to prove that $\Lambda = \Lambda^*$. Every point of Λ^* is the accumulation point of periodic points having arbitrarily large period. Hence $\Lambda^* \subset \Lambda$.

To show the reverse containment, suppose that $p \in \Lambda$. There exists a sequence of points $q_n \to p$ with the following property $f^1(q_n), ..., f^n(q_n)$ are all defined and distinct. Since Δ is dense, we can take a new sequence $\{q_n'\}$ of periodic points converging to p, and we can make $|q_n - q_n'|$ as small as we like. Making these distances sufficiently small, we guarantee that $f^1(q_n'), ..., f^n(q_n')$ are all distinct. This means that q_n' has period more than n.

Since X is compact, there are only finitely many periodic tiles having diameter greater than ϵ, for any $\epsilon > 0$. Hence, the size of the periodic tile containing q_n' necessarily converges to 0. But then every neighborhood of p intersects infinitely many periodic tiles. □

LEMMA 2.7. *When Δ is dense, Λ is closed.*

Proof: When Δ is dense, we have simply $\Lambda = X - \Delta$. Since Δ is open and X is closed, Λ is also closed. □

Remark: Λ is closed even when Δ is not dense. We leave this an an exercise for the interested reader.

2.6. Some Hyperbolic Geometry

This material is not used until Part 2 of the monograph. More details about hyperbolic geometry can be found in [**B**], [**BKS**], and [**S1**, §10,12].

The Mobius Group: $SL_2(\boldsymbol{C})$ is the group of 2×2 complex matrices of determinant 1. The matrix

$$M = \begin{bmatrix} a & b \\ c & d \end{bmatrix} \tag{2.9}$$

Acts on the Riemann sphere $\boldsymbol{C} \cup \infty$ by linear fractional (or Mobius) transformations

$$T_M(z) = \frac{az+b}{cz+d}. \tag{2.10}$$

This group action is compatible with matrix multiplication:

$$T_A \circ T_B = T_{AB}. \tag{2.11}$$

The two maps T_M and T_{-M} have the same action, and it is customary to work with the group $PSL_2(\boldsymbol{C}) = SL_2(\boldsymbol{C})/\pm I$. This group is often called the Mobius group. We will sometimes confound an element of $PSL_2(\boldsymbol{C})$ with the linear fractional transformation it determines.

Generalized Circles: A *generalized circle* in $\boldsymbol{C} \cup \infty$ is either a round circle or a topological circle of the form $L \cup \infty$ where $L \subset \boldsymbol{C}$ is a straight line. Mobius transformations map generalized circles to generalized circles, and preserve angles between them. See e.g. [**S1**, Theorem 10.1].

The subgroup $PSL_2(\boldsymbol{R})$ consists of those equivalence classes of real matrices. Elements of this subgroup preserve $\boldsymbol{R}\cup\infty$ and both the upper and lower half-planes. In particular, elements of $PSL_2(\boldsymbol{R})$ preserve the set of generalized circles which are either vertical lines or circles having the real axis as a diameter.

Hyperbolic Plane: The *hyperbolic plane* is the upper half plane in $\boldsymbol{C}$. We denote it by $\boldsymbol{H}^2$. We equip $\boldsymbol{H}^2$ with the Riemannian metric

$$\langle v, w\rangle_{x+iy} = \frac{v\cdot w}{y^2}. \tag{2.12}$$

The group $PSL_2(\boldsymbol{R})$ acts on $\boldsymbol{H}^2$ by Mobius transformations.

LEMMA 2.8. *$PSL_2(\boldsymbol{R})$ acts isometrically on $\boldsymbol{H}^2$.*

Proof: The claim is fairly obvious for the maps $z\to az+b$, and a direct calculation shows that the element $z\to -1/z$ also has this property. All other elements of $PSL_2(\boldsymbol{R})$, interpreted as linear fractional transformations, are compositions of the maps just mentioned. Compare [**S1**, §10.5]. □

The isometry group of $\boldsymbol{H}^2$ is generated by $PSL_2(\boldsymbol{R})$ and by (say) reflection in a vertical line. The *geodesics* (i.e. length minimizing paths) in $\boldsymbol{H}^2$ are either vertical rays or semicircles which meet $\boldsymbol{R}$ at right angles. We have already mentioned that $PSL_2(\boldsymbol{R})$ permutes these geodesics.

Ideal Triangles: Each geodesic in $\boldsymbol{H}^2$ has two endpoints in $\boldsymbol{R}\cup\infty$. We say that two geodesics are *asymptotic* if they have a common endpoint. For instance, the vertical geodesics $y=0$ and $y=1$ share ∞ as an endpoint. We say that an *ideal triangle* is the closed region bounded by three pairwise asymptotic geodesics. Each ideal triangle has 3 *ideal vertices*. One can find an element of $PSL_2(\boldsymbol{R})$ which maps any 3 distinct points in $\boldsymbol{R}\cup\infty$ to any other 3 distinct points in $\boldsymbol{R}\cup\infty$. For this reason, all ideal triangles are isometric. Ideal triangles are noncompact but have finite area equal to π.

Discrete Groups and Hyperbolic Surfaces: For us, a *hyperbolic surface* is a geodesically complete Riemannian surface that is locally isometric to $\boldsymbol{H}^2$. Hyperbolic surfaces are closely related to certain subgroup of $PSL_2(\boldsymbol{R})$, as we now explain.

A subgroup $\Gamma\subset PSL_2(\boldsymbol{R})$ is *discrete* if the identity element of Γ is not an accumulation point of Γ. An equivalent definition is that every convergent sequence in Γ is eventually constant. Yet another equivalent definition is that for any compact set $K\subset\boldsymbol{H}^2$, the set

$$\{\gamma\in\Gamma|\ \gamma(K)\cap K\neq\emptyset\} \tag{2.13}$$

is finite.

The group Γ is said to *act freely* if it never happens that $\gamma(p)=p$ for some $\gamma\in\Gamma$ and some $p\in\boldsymbol{H}^2$. When Γ is discrete and acts freely, the quotient $\boldsymbol{H}^2/\Gamma$ is a hyperbolic surface. Conversely, every hyperbolic surface arises this way. The proof just amounts to showing that the universal cover of the surface is $\boldsymbol{H}^2$ and that the deck group Γ is discrete and acts freely on $\boldsymbol{H}^2$.

2.7. Continued Fractions

Here we give a rapid introduction to the theory of continued fractions. See [**Da**] for a more complete exposition.

Gauss Map: The *Gauss map* is the map $G : (0,1) \to [0,1)$ defined by

$$G(x) = \frac{1}{x} - \mathrm{floor}\Big(\frac{1}{x}\Big). \tag{2.14}$$

We have $G(1/n) = 0$ when $n > 1$ is an integer. This map has an invariant measure, the *Gauss measure* $\mu = (1+x)^{-1}dx$. Here, the invariance means that $\mu(G^{-1}(S)) = \mu(S)$ for all measurable sets $S \subset [0,1]$. See [**BKS**, §1,4,5].

Continued Fractions: Given some $s \in (0,1)$ let $s_n = g^n(s)$. There are integers $a_1, ...$ so that

$$a_n = \mathrm{floor}\Big(\frac{1}{s_{n-1}}\Big), \qquad n = 1,2,3,... \tag{2.15}$$

The sequence $(0, a_1, a_2, a_3, ...)$ is called the continued fraction expansion of s. When $s > 1$ we set $a_0 = \mathrm{floor}(s)$ and then define $s = (a_0, a_1, a_2, ...)$ where $(0, a_1, a_2, ...)$ is the continued fraction expansion for $s - a_0$. The auxilliary sequence

$$a_0 + \frac{1}{a_1}, \qquad a_0 + \cfrac{1}{a_1 + \cfrac{1}{a_2}}, \qquad a_0 + \cfrac{1}{a_1 + \cfrac{1}{a_2 + \cfrac{1}{a_3}}}, \qquad \cdots \tag{2.16}$$

converges to s.

The Modular Group: The *Modular group* $PSL_2(\boldsymbol{Z}) \subset PSL_2(\boldsymbol{R})$ is the subgroup consisting of integer matrices. As we mentioned above, this group acts on $\boldsymbol{R} \cup \infty$ by linear fractional transformations The modular group is closely related to continued fractions. Suppose $s, s' \in (0,1)$ respectively have continued fractions $(0, n_0, n_1, ...)$ and $(0, n_0', n_1', ...)$. Then s and s' are in the same orbit of $SL_2(\boldsymbol{Z})$ if and only if $n_{s+k} = n'_{t+k}$ for some integers $s, t \geq 0$ of the same parity. This result is readily derivable from [**BKS**, Theorem 5.16], which treats the case when $s, s' > 1$. We will not need this result here, but it motivates a similar result involving our renormalization map R.

Recurrence formula: Suppose that p/q has continued fraction expansion

$$(a_0, a_1, ..., a_n).$$

We introduce numbers $p_{-2}, p_{-1}, p_0, ..., p_n$ and $q_{-2}, q_{-1}, q_0, ..., q_n$ such that

$$p_{-2} = 0, \quad p_{-1} = 1, \quad q_{-2} = 1, \quad q_{-1} = 0. \tag{2.17}$$

We then define the recurrence relation

$$p_k = a_k p_{k-1} + p_{k-2}, \qquad q_k = a_k q_{k-1} + q_{k-2}, \qquad k = 0, ..., n, \tag{2.18}$$

This recurrence relation gives $p = p_n$ and $q = q_n$. One shows inductively that

$$|p_k q_{k+1} - q_k p_{k+1}| = 1, \qquad \forall k. \tag{2.19}$$

This equation in turn implies that

(2.20)
$$\left|\frac{p_k}{q_k} - \frac{p_{k+1}}{q_{k+1}}\right| \leq \frac{1}{q_k^2}.$$

This inequality comes from the fact that $q_k \leq q_{k+1}$, as is easily seen from the recurrence relation.

Signed Continued Fractions: We define a *signed continued fraction* exactly like an ordinary continued fraction, except that we allow

(2.21)
$$a_k \in \boldsymbol{Z} - \{0, -1\}.$$

Thus, for instance, the S.C.F. $(0, 4, -3, 5, -7)$ means

$$\cfrac{1}{4 + \cfrac{1}{-3 + \cfrac{1}{5 + \cfrac{1}{-7}}}}$$

The recurrence relation described above works for S.C.F.s just as it does for ordinary continued fractions, and so do Equations 2.19 and 2.20. To get the basic fact $|q_k| \leq |q_{k+1}|$ we need to use the fact that $a_{k+1} \neq -1$.

2.8. Some Analysis

Hausdorff Convergence: Given a metric space M and two compact sets $S_1, S_2 \subset M$, one defines the *Hausdorff distance* $d(S_1, S_2)$ to be the infimal ϵ such that S_j is contained in the ϵ-neighborhood of S_{3-j} for $j = 1, 2$. By compactness, the infimum is actually realized. This puts a metric on the space of compact subsets of a metric space. We say that a sequence $\{S_n\}$ of closed (but not necessarily compact) subsets of M converges to S if, for every compact set K, we have $d(S_n \cap K, S \cap K) \to 0$ as $n \to \infty$. We call this the *Hausdorff topology* on the set of closed subsets of a metric space.

Hausdorff Dimension: In this section, we review some basic properties of the Hausdorff dimension. See [**F**] for more details.

Let M be a metric space. We let $|J|$ denote the diameter of a bounded subset $J \subset M$. Given a bounded subset $S \subset M$, and $s, \delta > 0$, we define

(2.22)
$$\mu(S, s, \delta) = \inf \sum |J_n|^s.$$

The infimum is taken over all countable covers of S by subsets $\{J_n\}$ such that $\text{diam}(J_n) < \delta$. Next, we define

(2.23)
$$\mu(S, s) = \lim_{\delta \to 0} \mu(S, s, \delta) \in [0, \infty].$$

This limit exists because $\mu(S, s, \delta)$ is a monotone function of δ. Note that $\mu(S, n) < \infty$ when $M = \boldsymbol{R}^n$. Usually we'll work in $\boldsymbol{R}^2$. Finally,

(2.24)
$$\dim(S) = \inf\{s |\ \mu(S, s) < \infty\}.$$

The number $\dim(S)$ is called the *Hausdorff dimension* of S.

We will mainly be concerned with upper bounds on Hausforff dimension.

LEMMA 2.9. *Suppose that there are infinitely many integers $m > 0$ such that S has a cover by at most m^D sets, all of which have diameter at most C/m. Here C is some constant that does not depend on m. Then* $\dim(S) \leq D$.

Proof: Choose and $s > D$. The existence of our covers tells us that $\mu(S, s, \delta) \to 0$ as $\delta \to 0$. Hence $\mu(X, s) = 0$. But then $\dim(S) \leq D$ by definition. □

CHAPTER 3

Multigraph PETs

3.1. The Abstract Construction

Elementary Members: We first describe some special members of **PET**, which we call *elementary*. Suppose that X and Y are convex polytopes, both fundamental domains for the same lattice L. We define partitions of X and Y respectively, by the equations

$$X = \bigcup_{V \in L} (Y - V) \cap X, \qquad Y = \bigcup_{V \in L} (X + V) \cap Y. \tag{3.1}$$

Both unions are finite because X and Y are compact. Note that translation by V carries $(Y - V) \cap X$ to $Y \cap (X + V)$. Hence, Equation 3.1 has the same form as Equation 2.4. Concretely, we have a PET $f : X \to Y$ defined by the equation $f(x) = x + V$ for all $x \in (V - Y) \cap X$. We call this map $X \to_L Y$.

Decorated Multigraphs: A *decoration* of a multigraph Γ is labelling of the vertices of Γ by convex polytopes and the edges by Euclidean lattices subject to the constraint that a vertex is incident to an edge iff the corresponding polytope is a fundamental domain for the corresponding lattice.

The Functor: Let $\mathbf{PATH}(\Gamma)$ denote the category whose objects are vertices of Γ and whose morphisms are paths connecting pairs of vertices of Γ. Once Γ has been decorated, we get a functor

$$\Phi : \mathbf{PATH}(\Gamma) \to \mathbf{PET}.$$

Φ maps each vertex to the corresponding polytope and each triple $x \to_\ell y$ to the elementary PET $X \to_L Y$. Here x and y are adjacent vertices connected by the edge ℓ and X, L, Y are the respective labels. Φ extends to all of $\mathbf{PATH}(\Gamma)$ so as to respect the composition laws.

Homotopy: Say that two elements of $\mathbf{PATH}(\Gamma)$ equivalent if there is an end-point fixing homotopy from the one path to the other. Φ maps equivalent paths to equivalent PETs. The reason is that the homotopy proceeds at each stage either by creating a path of the form $x \to_L \to y \to_L \to x$ or by deleting such a path, and the corresponding compositions of PETs are equivalent to the identity map.

Restriction to Loops: Suppose we fix a vertex basepoint $x \in \Gamma$ and let X be the corresponding polytope. Let $\mathbf{LOOP}(\Gamma, x)$ be the semigroup of loops in Γ which start and end at x. The functor above restricts to a map

$$\mathbf{LOOP}(\Gamma, x) \to \mathrm{PET}(X).$$

Here PET(X) is interpreted as the semigroup of PETs with domain X.

If we (re)define PET(X) as the set of equivalence classes of PETs on X, then PET(X) is a group, and we get the homormorphism

$$\pi_1(\Gamma, x) \to \mathrm{PET}(X)$$

mentioned in Equation 1.3.

Back to Earth: Here is a more concrete description of the whole construction. Imagine that we have a loop based at x, of the form

$$x = x_0, \ell_1, x_2, ..., \ell_{2n-1}, x_{2n} = x,$$

where ℓ_k is an edge connecting x_{j-1} to x_{j+1}. Let $X = X_0, ..., X_{2N} = X$ be the corresponding polytopes. and let $L_1, ..., L_{2n-1}$ be the corresponding lattices. Starting with a typical $p_0 \in X_0$, we choose the unique lattice vector $V_1 \in L_1$ such that $p_2 = p_0 + L_1 \in X_2$. We then choose the unique lattice vector $V_3 \in L_3$ such that $p_4 = p_2 + L_3 \in X_4$ and so on, until we reach $x_{2n} = f(p_0)$. The map f is our element of PET(X).

Remark: The reader might wonder why we went through so much abstraction to say something so simple. For one thing, we wanted to explain the functorial nature of our construction. For another thing, the approach above makes certain results automatic. For instance, one might wonder why the concretely defined map is really a PET. One way to see this is that what we get is a finite composition of the elementary members of **PET**, which is a category.

Double Lattice PETs As we mentioned in the introduction, we say that a *double lattice PET* is a multigraph PET corresponding to a decorated bigon. A *bigon* is a graph consisting of two vertices connected by two edges. In this case, $\pi_1(\Gamma, x) = \boldsymbol{Z}$, and the image in PET($X$) consists of powers of a single map.

3.2. The Reflection Lemma

The result in this section will give us a way to produce lots of examples of decorated multigraphs.

Let L be a lattice and suppose that

$$\beta = \{v_1, ..., v_m, w_1, ..., w_{n-m}\} \tag{3.2}$$

is a $\boldsymbol{Z}$-basis for L. By this we mean that L consists of the integer linear combination of vectors in the basis. Define $F_\beta = T_\beta([0,1]^n)$, where T_β is the linear transformation carrying the standard basis to β. Let $F = F_\beta$. Certainly F is a fundamental domain for L. However, we are going to show that F is also the fundamental domain for a different lattice.

Let Π_V denote the k-dimensional linear subspace spanned by $v_1, ..., v_m$. Let R_V denote isometric reflection in Π_V. The lattice $R_V(L)$ is spanned by the vectors

$$R_V(\beta) = \{v_1, ..., v_m, R_V(w_1), ..., R_V(w_{n-m})\} \tag{3.3}$$

LEMMA 3.1 (Reflection). *F is a fundamental domain for $R_V(L)$.*

Proof: Let $L' = R_V(L)$. Note that $\boldsymbol{R}^n/L$ and $\boldsymbol{R}^n/L'$ have the same volume. Also $\boldsymbol{R}^n/L$ and F have the same volume. Therefore, $\boldsymbol{R}^n/L'$ and F have the same volume. To finish the proof, we have to take an arbitrary point $p \in \boldsymbol{R}^n$ and show that there is some vector $V \in L'$ such that $p + V \in F$.

Let H be the subspace spanned by $v_1, ..., v_m$. The map R_V preserves the orthogonal complement $H^\perp$ and acts on $H^\perp$ as reflection through the origin. Let $\pi : \boldsymbol{R}^n \to H^\perp$ be orthogonal projection. The kernel of π is exactly H. Moreover, $\pi(F)$ is a fundamental domain for $\pi(L)$ in $H^\perp$. Since R_V acts as reflection through the origin on $H^\perp$, we have $\pi(L) = -\pi(L')$. At the same time $\pi(L')$ is a lattice in $H^\perp$, so that $-\pi(L') = \pi(L')$. Hence

$$\pi(L) = \pi(L'). \tag{3.4}$$

But then $\pi(F)$ is a fundamental domain for $\pi(L')$. Therefore, there is some $V' \in L'$ such that the translate H' of H, through $p + V'$, intersects F.

Since F is a parallelotope which intersects H in a parallelotope, the intersection $H' \cap F$ is isometric to $H \cap F$, and hence is a fundamental domain for the lattice $L'' = \boldsymbol{Z}(v_1, ..., v_m)$. Hence, there is some $V_2 \in L''$ such that $p + V' + V'' \in F$. But $V = V' + V'' \in L'$. So, $p + V \in F$ for some $V \in L'$. This proves what we want. □

3.3. Constructing Multigraph PETs

Nice Groups: An *isometric involution* is an order 2 isometry of $\boldsymbol{R}^n$ which fixes the origin. We call G a *nice group* if G is a finite group of isometries generated by isometric involutions, and no odd product of the generators is trivial. When G is nice, we can speak of *odd* (respectively *even*) elements of G as being those which are a product of an odd (respectively even) number of generators.

Adapted Parallelotopes: Say that a *marked parallelotope* is a parallelotope P with a distinguished vertex p. The pair (P, p) naturally determines a basis in $\boldsymbol{R}^n$. The basis consists of the vectors $w - p$, where w is a vertex of P adjacent to p. Suppose that G is a nice group, which is generated by isometric involutions $I_1, ..., I_k$. We say that a marked parallelotope (P, p) is *adapted* to G if the fixed point set of each I_j is spanned by some subset S_j of the basis vectors associated to (P, p). We call S_j the *list associated to* I_j.

The Main Construction: If the origin is a vertex for P, we let (P) be the marked parallelotope $(P, 0)$. Suppose P is adapted to a nice group G. Let Γ be the following graph. The vertices of Γ are the odd elements of G. The edges connect the pairs (g, gI_jI_k), where g is an odd element of G. We decorate Γ as follows:

- The vertex g of Γ is labelled the translate of $g(P)$ that is centered at the origin.
- The edge e connecting g to gI_jI_k is labelled by the lattice L generated by the basis β associated to $(h(P))$.

We now verify the compatibility condition. Let β' be the basis associated to $(g(P))$. Let S_j be the list associated to I_j. Let $v_1, ..., v_m$ be the elements of $g(S_j)$. Let R_V be isometric reflection in the subspace spanned by $v_1, ..., v_n$. We have $\beta' = R_V(\beta)$, because

$$h(P) = R_V(g(P)), \qquad R_V = gI_jg^{-1}. \tag{3.5}$$

By the Reflection Lemma, $g(P)$, the parallelotope associated to g, is a fundamental domain for $\boldsymbol{Z}[\beta]$, the lattice associated to e.

Since the compatibility condition holds for all pairs (g,e), we get a decorated multigraph whose underlying graph is Γ.

3.4. Planar Examples

Here are the basic examples in the plane. We call these examples the *dihedral PETs*.

- Let $G = D_{2n}$, the dihedral group of order $2n$. We generate G by reflections I_1 and I_2, where I_1 is reflection in the x-axis and I_2 is reflection in the line L_n making an angle of π/n with the x-axis.
- Let P be the parallelogram having vertices
$$(0,0), \qquad (2,0), \qquad w_s, \qquad w_s + (2,0).$$
Here w_s is the point in L_n, depending on the parameter s. The distinguished vertex is $p = (0,0)$. The corresponding basis β is given by $(2,0)$ and w_s.
- The graph Γ is an n-cycle. Our decoration convention is to translate all the parallograms so that they are centered at the origin.

When n is even, the element $\iota = (I_1 I_2)^{n/2}$ is the reflection in the origin. In this case, our decoration of Γ commutes with the corresponding order 2 automorphism of Γ, which we also call ι. The quotient Γ/ι is a decorated $(n/2)$-cycle.

In the case $n = 1$, there is nothing to define. The case $n = 2$ leads to a decorated monogon, and the corresponding PET is the identity map. The case $n = 4$ is is a double lattice PET. In this case, we take $w_s = (2s, 2s)$, and we recover the octagonal PETs. We think that, in general, the cases $n = 3, 4, 5, ...$ correspond to outer billiards on semi-regular $(2n)$-gons in the same way that the case $n = 4$ corresponds to outer billiards on semi-regular octagons.

We have not yet investigated the dihedral PETs beyond the case of the octagonal PETs. Also, we have not spent much time investigating even the variants of the octagonal PET in which the parallelograms are not centered at the origin. Some preliminary computer investigation suggests that (for irrational parameters) such such systems tend to have dense orbits, or at least orbits which are dense in open subsets of the plane. Thanks to this experimentally observed phenomena, we think that the convention of translating the parallelotopes so that they are centered at the origin is, in general, a good idea.

3.5. Three Dimensional Examples

One can give a family of examples based on spherical triangle reflection groups. The symbol (a,b,c) denotes the spherical triangle τ with angles π/a, π/b, and π/c. One must have the condition $1/a + 1/b + 1/c > 1$, and to get a finite reflection group we want $a,b,c \in \boldsymbol{Z}$. This leaves

- $(2,3,3)$: tetrahedral case.
- $(2,3,4)$: cube/octahedron case.
- $(2,3,5)$: icosahedron/dodecahedron case.
- $(2,2,n)$ for $n = 2,3,4,5...$: dihedral case.

In each of the first three cases, the group G generated by reflections in the sides of the triangle is the full symmetry group of the corresponding platonic solid(s).

To produce a decorated multigraph, we choose vectors w_1, w_2, w_3 such that w_i is a positive multiple of the ith vertex of the triangle τ. The parallelapiped P is such that $p = (0,0,0)$ is the distinguished vertex, and w_1, w_2, w_3 are the vertices adjacent to p. By construction (P,p) is adapted to G. Once again (though we have no especially good reason to do this) we use the convention that the parallelopipeds labelling the vertices of the graph Γ are centered at the origin.

The case $(2,2,2)$ is not interesting at all; all the vertices get the same label, namely a translated copy of P, and all the edges get the same label, namely the lattices $\boldsymbol{Z}(w_1, w_2, w_3)$. The case $(2,2,n)$ for $n > 2$ seems similar in spirit to the dihedral PETs constructed in the previous section.

In the cases $(2,3,4)$ and $(2,3,5)$, some element of G is reflection in the origin, and so the opposite vertices and edges of Γ get the same labels. Here, again, it probably makes sense to quotient out by this symmetry. The case $(2,3,3)$ does not have this symmetry.

In all of these cases, the PETs we get depend on the choice of vectors w_1, w_2, w_3. The choice $\lambda w_1, \lambda w_2, \lambda w_3$ leads to the same systems modulo scale, so really we have a 2-parameter family of decorated multigraph PETs for each choice of spherical triangle. I would guess that these families are quite rich, and perhaps some version of Theorem 1.10 is true, with respect to lattices in $PSL_3(\boldsymbol{R})$ rather than a lattice in $PSL_2(\boldsymbol{R})$.

3.6. Higher Dimensional Generalizations

We could generalize the constructions in the previous section by considering the reflection symmetry groups of higher dimensional regular solids. Here we will generalize in a different direction. The 2-dimensional cases of our construction here are the octagonal PETs.

Let $G = D_4$ act on the plane, as in our description of the octagonal PET above. G also acts on $\boldsymbol{R}^{2n} = (\boldsymbol{R}^2)^n$ by acting on each factor of $\boldsymbol{R}^2$ in the same way that it acts on $\boldsymbol{R}^2$ itself. The two generators I_1 and I_2 of G both fix n-dimensional subspaces.

We can form a parallelotope P by choosing a basis $\beta_1 = \{v_1, ..., v_n\}$ for the fixed subspace of I_1, and a basis $\beta_2 = \{w_1, ..., w_n\}$ for the fixed subspace of I_2. Then, P is the parallelotope such that $p = (0, ..., 0)$ is the distinguished vertex and the points in $\beta_1 \cup \beta_2$ are the adjacent vertices. By construction (P,p) is adapted to G. The resulting decorated multigraph Γ is a 4-cycle, but again the opposite vertices and edges get the same labels. So, taking the quotient, we get a decorated bigon. This gives us a double lattice PET for each choice of basis.

Let $M \in GL_n(\boldsymbol{R})$ denote any invertible real $n \times n$ matrix. We define T_M : $\boldsymbol{R}^{2n} \to \boldsymbol{R}^{2n}$ as follows. We identify $(\boldsymbol{R}^2)^n$ with $\boldsymbol{C}^n$ in the usual way and then let M act on $\boldsymbol{C}^n$. Under this identification, the fixed subspaces of both I_1 and I_2 are totally real. The map T_M preserves these subspaces and indeed commutes with the action of G. Hence, if we replace β_1 and β_2 with $T_M(\beta_1)$ and $T_M(\beta_2)$ we get conjugate PETs. So, up to conjugacy, we might as well consider systems in which β_1 is the standard basis on $\boldsymbol{R}^n = \text{Fix}(I_1)$. So, the space of inequivalent PETs produced by our construction is naturally indexed by the set of bases β_2, which is just a copy of the matrix group $GL_n(\boldsymbol{R})$. It is worth noting that the systems corresponding to the subgroup $GL_n(\boldsymbol{Q})$ have a nice property.

LEMMA 3.2. *For any system indexed by a member of $GL_n(\boldsymbol{Q})$, all the orbits are periodic.*

Proof: We can dilate the whole picture so that both bases β_1 and β_2 consist entirely of vectors having integer coordinates. Letting (X, f) denote the associated PET, we observe that all the points in an orbit of f differ from each other by integer vectors. Moreover, such orbits are bounded. Hence, they are finite. □

CHAPTER 4

The Alternating Grid System

4.1. Basic Definitions

In this chapter we will introduce the alternating grid systems, and then prove Theorems 1.12 and 1.13.

We fix some $s \in (0, \infty)$. Let $G_{s,0}$ denote the infinite grid of squares in the plane such that the sides of the squares have side length s and are parallel to the coordinate axes. We pin down $G_{s,0}$ by insisting that $(0,0)$ is a vertex of a square in $G_{s,0}$. Let $T_{s,0}$ denote the map which rotates each square of $G_{s,0}$ by $\pi/2$ radians counterclockwise. That is, T_s performs a counterclockwise quarter turn to each square in $G_{s,0}$.

As we remarked in the introduction, the map [1]

$$F_{s,0} = T_{s,0}T_{1,0}T_{s,0}T_{1,0} \tag{4.1}$$

seems to be the most interesting word in the group generated by these elements. The map $F_{s,0}$ is a PET whose domain is $\boldsymbol{R}^2$. The periodic tiles of $F_{s,0}$ are necessarily rectangles.

The map $F_{s,0}$ seems to have quite interesting dynamics. What we notice experimentally is that almost every orbit of $F_{s,0}$ is periodic and that $F_{s,0}$ has erratic orbits. An *erratic orbit* is an unbounded orbit which enters a compact subset of the plane infinitely often.

We also notice that there seems to be a highly structured collection of periodic tiles

$$P_{1,1}, ..., P_{1,n_1}, P_{2,1}, ..., P_{2,n_2}, ... \tag{4.2}$$

- The collection converges to the origin as $i \to \infty$.
- The tiles $P_{i1}, ..., P_{in_i}$ all have the same size.
- The sequence $(0, n_1, n_2, n_3, ...)$ is the continued fraction expansion of s. See §2.7 for an account of continued fractions.
- As $i \to \infty$ the diameter of the orbit of the ith tile tends to ∞.
- For $i = 2, 4, 6...$ the tiles P_{ij} all share an edge with the x-axis.
- For $i = 3, 5, 7...$ the tiles P_{ij} all share an edge with the y-axis.

Call this the *continued fraction conjecture*.

Figure 4.1 shows how the unit square is tiled by the periodic tiles for $s = 36/61 = (0, 1, 1, 2, 3, 1, 2)$. The sequence denotes the continued fraction expansions of s. The tiles involved in the continued fraction conjecture have been shaded.

[1]The composition means "first apply $T_{1,0}$, then apply $T_{s,0}$, etc."

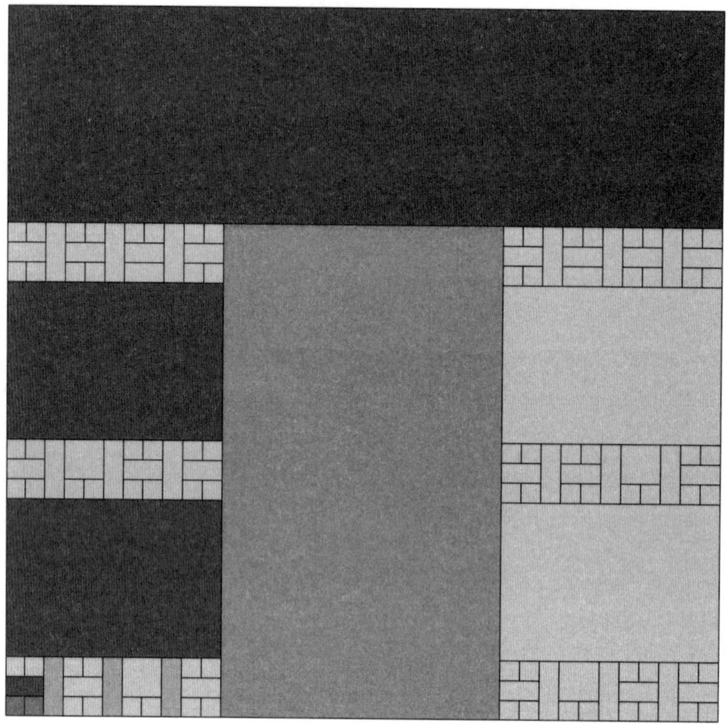

Figure 4.1: Tiling of the unit square for $s = 36/61$.

We will not really study the dynamics of the alternating grid systems in this monograph, and in particular we will not make any progress towards proving the continued fraction conjecture – we don't know how to prove it. Our main purpose here is to show the connection between the alternating grid system, double lattice PETs, and the octahedral PETs. However, our interest in proving the continued fraction conjecture lead to our discovery of the octagonal PETs. The orbits involved in the continued fraction conjecture are both intricate and wide-ranging. In order to bring them all into view, so to speak, it seemed like a good idea to compactify the system, and then the invariant slices of the compactifications turned out to be the octagonal PETs.

4.2. Compactifying the Generators

We considered the map $F_{s,0}$ above because we wanted to highlight the continued fraction conjecture. For the purposes of describing the compactification, we find it more natural to reposition the grids. When s is irrational, any repositioning of the grids leads to the same compactification. Let G_s be the translate of $G_{s,0}$ such that the origin is the center of one of the squares. In terms of the notation in §1.8, we have $G_s = G_{s,(s/2,s/2)}$. We define T_s as above, relative to G_s, and then $F_s = T_s T_1 T_s T_1$.

Define the flat torus

$$\boldsymbol{T}^n = \boldsymbol{R}^n / \boldsymbol{Z}^n. \tag{4.3}$$

When we want to give coordinates to points in $\boldsymbol{T}^n$, we think of it as the cube $[-1/2,1/2]^n$ with opposite sides identified. Let O_n denote the origin in $\boldsymbol{R}^n$. Note that O_n is the center of our cube.

We sometimes suppress the parameter s in our notation. For convenience, we take s to be irrational. We define $\Psi : \boldsymbol{R}^2 \to \boldsymbol{T}^4$ by the map

$$\Psi(x,y) = \left(x, y, \frac{x}{s}, \frac{y}{s}\right) \mod \boldsymbol{Z}^4. \tag{4.4}$$

Observe that Ψ maps the fixed points of T_1 and T_s respectively to dense subsets of $O_2 \times \boldsymbol{T}^2$ and $\boldsymbol{T}^2 \times O^2$ respectively.

We introduce matrices

$$M_1 = \begin{bmatrix} 0 & -1 & 0 & 0 \\ 1 & 0 & 0 & 0 \\ -1/s & -1/s & 1 & 0 \\ 1/s & -1/s & 0 & 1 \end{bmatrix}, \qquad M_s = \begin{bmatrix} 1 & 0 & -s & -s \\ 0 & 1 & s & -s \\ 0 & 0 & 0 & -1 \\ 0 & 0 & 1 & 0 \end{bmatrix}. \tag{4.5}$$

We have the relations

$$M_s^4 = M_1^4 = (M_s M_1)^2 = (M_s M_1)^2 = I, \tag{4.6}$$

where I is the identity matrix. M_1 and M_s respectively act as the identity on the subspaces $\boldsymbol{R}^2 \times O_2$ and $O_2 \times \boldsymbol{R}^2$. For this reason, the maps

$$\widehat{T}_a(p) = M_a(p) \mod \boldsymbol{Z}^4, \qquad a \in \{1, s\} \tag{4.7}$$

are well defined respectively on the spaces

$$\boldsymbol{T}_1^4 = \left(-\frac{1}{2}, \frac{1}{2}\right)^2 \times \boldsymbol{T}^2, \qquad \boldsymbol{T}_s^4 = \boldsymbol{T}^2 \times \left(-\frac{1}{2}, \frac{1}{2}\right)^2. \tag{4.8}$$

However, $\widehat{T}_a$ is not defined on the complement $\boldsymbol{T}^4 - T_a^4$ which, in both cases, is a union of 4 intersecting copies of $\boldsymbol{T}^3$.

LEMMA 4.1. *$\widehat{T}_1 \circ \Psi = \Psi \circ T_1$.*

Proof: Let $p \in \boldsymbol{R}^2$ be in the domain of T_1. We can write

$$p = (x_1, x_2) + (n_2, n_2), \qquad x_j \in (-1/2, 1/2), \qquad n_j \in \boldsymbol{Z}. \tag{4.9}$$

We have

$$T_1(p) = (-x_2, x_1) + (n_1, n_2), \tag{4.10}$$

$$\Psi(p) = \left(x_1, x_2, \frac{x_1}{s}, \frac{x_2}{s}\right) + (0, 0, n_1', n_2') \in \boldsymbol{T}_1^4, \tag{4.11}$$

$$\Psi \circ T_1(p) = \left(-x_2, x_1, -\frac{x_2}{s}, \frac{x_1}{s}\right) + (0, 0, n_1', n_2') \in \boldsymbol{T}_1^4. \tag{4.12}$$

Here n_j' is the equivalence class of n_j/s mod $\boldsymbol{Z}$. The addition makes sense in $\boldsymbol{T}_1^4$ because we are only adding the last two coordinates. A direct matrix calculation shows that $M_1 \circ \Psi(p)$ equals the right hand side of Equation 4.12. Interpreting $M_1 \circ \Psi(p)$ as a point in $\boldsymbol{T}_1^4$, we get the desired equality. □

LEMMA 4.2. *$\widehat{T}_s \circ \Psi = \Psi \circ T_s$.*

Proof: Let $p \in \boldsymbol{R}^2$ be in the domain of T_s. We can write

$$p = (sx_1, sx_2) + (sn_2, sn_2), \qquad x_j \in (-1/2, 1/2), \qquad n_j \in \boldsymbol{Z}. \tag{4.13}$$

This time, we have

$$T_s(p) = (-sx_2, sx_1) + (sn_1, sn_2) \in \boldsymbol{T}^4_s, \tag{4.14}$$

$$\Psi(p) = \Big(sx_1, sx_2, x_1, x_2\Big) + (n_1', n_2', 0, 0) \in \boldsymbol{T}^4_s, \tag{4.15}$$

$$\Psi \circ T_s(p) = \Big(- sx_2, sx_1, -x_2, x_1\Big) + (n_1', n_2', 0, 0) \in \boldsymbol{T}^4_2. \tag{4.16}$$

Here n_j' is the equivalence class of sn_j mod $\boldsymbol{Z}$. A direct matrix calculation shows that $M_2 \circ \Psi(p)$ equals the right hand side of Equation 4.12, and this completes the proof as in the previous case. □

4.3. The PET Structure

Lemmas 4.1 and 4.2 combine to show that Ψ is a semi-conjugacy between the group $\langle T_1, T_s\rangle$ acting on $\boldsymbol{R}^2$ and the group $\langle \widehat{T}_1, \widehat{T}_s\rangle$ acting on $\boldsymbol{T}^4$. In both cases, as usual, the elements of the groups are only defined almost everywhere – or, more precisely, on the complement of some codimension 1 set. In particular, if we define $\widehat{F} = \widehat{T}_s\widehat{T}_1\widehat{T}_s\widehat{T}_1$, then $\Psi \circ F = \widehat{F} \circ \Psi$. Hence $(\boldsymbol{T}^4, \widehat{F})$ is a compactification for $(\boldsymbol{R}^2, F)$, as in Theorem 1.12.

The maps $\widehat{T}_1$ and $\widehat{T}_s$ are locally affine on $\boldsymbol{T}^4_1$ and $\boldsymbol{T}^4_s$ respectively, though they do not extend continuously to all of $\boldsymbol{T}^4$. The linear parts of $\widehat{T}_1$ and $\widehat{T}_s$ are M_1 and M_s respectively. The map $\widehat{F}$ is the composition of these locally affine maps. The linear part of $\widehat{F}$ is $M_sM_1M_sM_1$, which is the identity! Hence $\widehat{F}$ is a piecewise translation. To complete the proof of Theorem 1.12, we will recognize $\widehat{F}$ as a double lattice PET.

LEMMA 4.3. *$\widehat{T}_s\widehat{T}_1$ is a well-defined, and locally affine, on a dense subset of $\boldsymbol{T}^4$ which is isometric to the interior of a convex parallelotope.*

Proof: The domain for $\widehat{T}_s\widehat{T}_1$ is the open dense set

$$P = \boldsymbol{T}^4_1 \cap T_1^{-1}(\boldsymbol{T}^4_s) \subset \boldsymbol{T}^4. \tag{4.17}$$

To understand P, we consider the picture in the universal cover. Define

$$\widetilde{P} = \boldsymbol{R}^4_1 \cap M_1^{-1}(\boldsymbol{R}^4_s),$$

where $\boldsymbol{R}^4_1$ (respectively $\boldsymbol{R}^4_s$) is the universal cover of $\boldsymbol{T}^4_1$ (respectively $\boldsymbol{T}^4_s$.) It is the set of points (x_1, x_2, x_3, x_4) so that x_1, x_2 (respectively x_3, x_4) do not have the form $n/2$ where n is an odd integer.

$\boldsymbol{R}^4_1$ and $M_1^{-1}(\boldsymbol{R}^4_s)$ are both the complements of a pair of parallel families of hyperplanes. Hence $\widetilde{P}$ is the complement of 4 infinite parallel families of hyperplanes. The corresponding normals are

$$e_1, \qquad e_2, \qquad M_1^t(e_3), \qquad M_1^t(e_4).$$

Up to sign, these are just the rows of M_1. The normals form a basis, so that $\widetilde{P}$ is a $\boldsymbol{Z}^4$ invariant infinite union of parallelotopes.

Since $P = \widetilde{P} \bmod \boldsymbol{Z}^4$ and (from Equation 4.17) ∂P is contained in a union of 8 flat subspaces of $\boldsymbol{T}^4$ we must have that P is isometric to an open parallelotope. □

Let $P_1 = P$, the parallelotope from the previous result. Define

$$P_2 = \widehat{T}_s\widehat{T}_1(P_1) \subset \boldsymbol{T}^4. \tag{4.18}$$

Since $\widehat{T}_s\widehat{T}_1$ is locally affine on P_1, the set P_2 is also isometric to the interior of a parallelotope. Since the linear part of $\widehat{T}_s\widehat{T}_1$, namely M_sM_1, is an involution, we see that P_1 and P_2 have the same volume. In particular, P_2 is also dense in $\boldsymbol{T}^4$.

By construction, reflection in the origin preserves P_1. Since $\widehat{T}_s\widehat{T}_1$ fixes the origin, reflection in the origin preserves P_2. For $j = 1, 2$, let X_j denote the lift of P_j to $\boldsymbol{R}^4$ which is centered at the origin. Let $\beta_j : X_j \to P_j$ denote the isometry which amounts to reduction mod $\boldsymbol{Z}^4$. Lemma 4.5 below characterizes X_1 and X_2.

Define

$$I = M_sM_1, \qquad L_1 = \boldsymbol{Z}^4, \qquad L_2 = I(\boldsymbol{Z}^4). \tag{4.19}$$

By construction I is an involution which swaps X_1 and X_2 and also swaps L_1 and L_2. Also by construction X_1 and X_2 are both fundamental domains for $\boldsymbol{Z}^4$. Given the symmetry under I, we see that X_1 and X_2 are fundamental domains for L_2. Therefore, the quadruple (X_1, X_2, L_1, L_2) defines a double lattice PET, $D : X_1 \to X_1$.

LEMMA 4.4. *$\beta_1^{-1} \circ \widehat{F} \circ \beta_1 = D$.*

Proof: Note that we are trying to prove that two different maps on X_1 are the same. Recalling that $F = \widehat{T}_s\widehat{T}_1\widehat{T}_s\widehat{T}_1$, we write

$$\beta_1^{-1}\widehat{F}\beta_1 = A \circ B, \qquad A = \beta_1^{-1}\widehat{T}_s\widehat{T}_1\beta_2, \qquad B = \beta_2^{-2}\widehat{T}_s\widehat{T}_1\beta_1. \tag{4.20}$$

The restriction $B|_{X_1}$ is linear, and the linear part of $B|_{X_1}$ is I. Therefore $B = I$ on X_1.

Let μ_{ij} denote the map from $\boldsymbol{R}^4$ to X_j which simply translates by vectors of L_i. We have

$$\beta_2 = \beta_1\mu_{11}, \qquad \beta_1^{-1} = \mu_{11}\beta_2^{-1}. \tag{4.21}$$

The first equation implies the second one. Therefore

$$A = (\beta_1^{-1}\beta_2)B(\beta_1^{-1}\beta_2) = \mu_{11} \circ I \circ \mu_{11}. \tag{4.22}$$

But then we have

$$\beta_1^{-1} \circ \widehat{F} \circ \beta_1 = \mu_{11} \circ I \circ \mu_{11} \circ I = \mu_{11} \circ \mu_{22} = D. \tag{4.23}$$

This completes the proof. □

4.4. Characterizing the PET

We keep the notation from the previous section. Here we will describe (X_1, X_2, L_1, L_2) more explicitly. Let

$$Q = (-1/2, 1/2)^4. \tag{4.24}$$

LEMMA 4.5. *$X_1 = M_1^{-1}(Q)$ and $X_2 = M_s(Q)$.*

Proof: Since $X_2 = M_s M_1(X_1)$, it suffices to prove the first equation. Let's look a little bit more closely at the proof of Lemma 4.3. Say that a *half-odd number* is a number of the form $n/2$ where n is an odd integer. The set $\boldsymbol{R}_1^4$ consists of those points v such that $v \cdot e_1$ and $v \cdot e_2$ are not half-odd numbers. The set $\boldsymbol{R}_2^4$ consists of vectors v such that $v \cdot M_1^t(e_3)$ and $v \cdot M_1^3(e_4)$ are not half-odd numbers. We can equally well say that $\boldsymbol{R}_1^4$ consists of those vectors v such that $v \cdot M_1^t(e_1)$ and $v \cdot M_1^t(e_2)$ are not half-odd numbers. Hence, X consists of those vectors v such that $v \cdot M_1^t(e_j) \in (-1/2, 1/2)$. But then

$$v \cdot M_1^t(e_j) = M_1(v) \cdot e_j \in (-1/2, 1/2).$$

In other words, all coordinates of $M_1(v)$ lie in $(-1/2, 1/2)$. $\square$

We have already mentioned that $L_1 = \boldsymbol{Z}^4$ and

$$L_2 = M_s M_1(\boldsymbol{Z}_4). \tag{4.25}$$

So, L_2 is the $\boldsymbol{Z}$-span of the columns of

$$I := M_s M_1 = \begin{bmatrix} 0 & 1 & -s & -s \\ -1 & 0 & s & -s \\ -1/s & 1/s & 0 & -1 \\ -1/s & -1/s & 1 & 0 \end{bmatrix} \tag{4.26}$$

Note that I is an involution.

Remark: It is worth pointing out that we might not have the most symmetric picture. We are free to conjugate the whole picture by some linear transformation γ. Perhaps the new quadruple $(\gamma(X_1), \gamma(X_2), \gamma(L_1), \gamma(L_2))$ will exhibit more symmetry. This approach is similar in spirit to what is done in [**AKT**] for their main example.

4.5. A More Symmetric Picture

4.5.1. Canonical Coordinates. Let us conjugate the picture by the matrix

$$\gamma = \begin{bmatrix} 0 & 0 & s & s \\ 0 & -2 & s & s \\ 0 & 0 & s & -s \\ -2 & 0 & s & -s \end{bmatrix} \tag{4.27}$$

We replace the objects X_j and L_j with $\chi_j = \gamma(X_j)$ and $\Lambda_j = \gamma(L_j)$. Here Λ_1 is the $\boldsymbol{Z}$-span of the columns of γ and Λ_2 is the $\boldsymbol{Z}$-span of the columns of

$$\gamma I = \begin{bmatrix} -2 & 0 & s & -s \\ 0 & 0 & -s & s \\ 0 & 2 & -s & -s \\ 0 & 0 & s & s \end{bmatrix} \tag{4.28}$$

$\chi_1 = \gamma M_1^{-1}(Q)$, and

$$\gamma M_1^{-1} = \begin{bmatrix} -2 & 0 & s & s \\ 0 & 0 & s & s \\ 0 & 2 & s & -s \\ 0 & 0 & s & -s \end{bmatrix} \tag{4.29}$$

$\chi_2 = \gamma M_s(Q)$, and

$$\gamma M_s = \begin{bmatrix} 0 & 0 & s & -s \\ 0 & -2 & -s & s \\ 0 & 0 & -s & -s \\ -2 & 0 & s & s \end{bmatrix} \tag{4.30}$$

The two reflections

$$\rho_1(x_1,x_2,x_3,x_4) = (-x_1,x_2,-x_3,x_4), \qquad \rho_2(x_1,x_2,x_3,x_4) = (x_2,x_1,x_4,x_3) \tag{4.31}$$

generate the order 8 dihedral group G, and (χ_1, p) is adapted to G, when p is chosen so that the basis determined by (χ_1, p) is given by the columns of the matrix in Equation 4.29. For each pair (i,j), there is some element of G which relates χ_i and Λ_j as in the Reflection Lemma. So, our compactification fits into the examples discussed in §3.6.

Remark: When $s = 1$, we have $\Lambda_1 = \Lambda_2 = E_4$, the famous lattice.

4.5.2. The Double Foliation. The composition

$$\Upsilon = \gamma \circ \beta \circ \Psi : \boldsymbol{R}^2 \to \chi_1. \tag{4.32}$$

conjugates the alternating grid map to the PET dynamics $\widehat{F} : \chi_1 \to \chi_1$. For almost all s, the map Υ is injective. The image

$$\mathcal{R}_1 = \Upsilon(\boldsymbol{R}^2) \tag{4.33}$$

is an invariant leaf of the linear foliation whose tangent planes are spanned by the vectors

$$\gamma(s^{-1},0,1,0) = (1,1,1,-1), \qquad \gamma(0,-s^{-1},0,-1) = (-1,1,1,1). \tag{4.34}$$

When Υ is injective, the restriction $\widehat{F}|\mathcal{R}_1$ can be identified with the alternating grid map.

LEMMA 4.6. *The involution* $\iota(x_1,x_2,x_3,x_4) = (x_1,x_2,-x_3,-x_4)$ *preserves* χ_j *and* Λ_j *for* $j = 1,2$.

Proof: Up to sign, ι fixes the two first columns of γ and swaps the last two columns. Hence $\iota(\Lambda_1) = \Lambda_1$. The same holds for γI. Hence $\iota(\Lambda_2) = \Lambda_2$.

Similarly, a calculation shows that, up to sign, the columns of $\iota\gamma M_1^{-1}$ are permutations of the columns of γM_1^{-1}. For this reason,

$$\iota(\chi_1) = \iota\gamma M_1^{-1}(Q) = \gamma M_1^{-1}(Q) = \chi_1.$$

A similar argument shows that $\iota(\chi_2) = \chi_2$. □

The image

$$\mathcal{R}_2 = \iota(\mathcal{R}_1). \tag{4.35}$$

is an invariant leaf of the linear foliation whose tangent planes are spanned by

$$(1,1,-1,1), \qquad (1,-1,1,1). \tag{4.36}$$

For almost all s, the leaves $\mathcal{R}_1$ and $\mathcal{R}_2$ are dense in their respective foliations. Hence, $\widehat{F}$ preserves every leaf of the corresponding foliations. In particular, $\widehat{F}$ preserves every translate of $\mathcal{R}_1$. We will use this fact when we discuss unbounded orbits below.

4.5.3. The Octagonal PETs. Now we consider the picture in the invariant subspaces of ι, namely the planes $\boldsymbol{R}^2_{12}$ and $\boldsymbol{R}^2_{34}$. Let Π be either of these planes. We identify Π with $\boldsymbol{R}^2$ in the obvious way – just drop off the zero coordinates. Referring to Figure 1.1, it is easy to check that

$$F_j = \Pi \cap X_j, \qquad L_j = \Pi \cap \Lambda_j \tag{4.37}$$

for $j = 1, 2$. Just take one representative example, $\boldsymbol{R}^2_{12} \cap \Lambda_1$ is the $\boldsymbol{Z}$-span of $(0,2,0,0)$ and $(2s, 2s, 0, 0)$, exactly as is L_1 in Figure 1.1.

In short, we see the octagonal PET at parameter s in both slices. This completes the proof of Theorem 1.13.

4.6. Unbounded Orbits

Let s be an irrational parameter. From the description above, the foliation containing $\mathcal{R}_1$ is transverse to

$$F_1 = \boldsymbol{R}^2 \cap \chi, \tag{4.38}$$

the domain of the octagonal PET. Let $F_{s,z}$ be the alternating grid map in Theorem 1.14. Let Λ'_s be the aperiodic set for the octagonal PET at parameter s. We know that Λ'_s is nonempty, by Statement 4 of Theorem 1.5.

LEMMA 4.7. *Suppose $F_{s,z}$ has unbounded orbits for some $z \in \boldsymbol{C}$.*

Proof: We have already remarked that $\widehat{F}_s$ preserves every translate of $\mathcal{R}_1$. In particular, there is a translate $\boldsymbol{R}^2_{s,z}$ of $\mathcal{R}_1$ such that $\widehat{F}_s$ preserves $\boldsymbol{R}^2_{s,z}$ and acts there as $F_{s,z}$. This translate is again transverse to F_1. We can choose z such that $\boldsymbol{R}^2_{s,z} \cap F_1$ contains a point p having infinite orbit O_p. By transversality, and the piecewise linear nature of the embedding Υ, each bounded subset of $\boldsymbol{R}^2$ includes into a subset of $\boldsymbol{R}^2_{s,z}$ which intersects F_1 in a finite set. If O_p is bounded in $\boldsymbol{R}^2$, then O_p is a finite subset of F_1, and this would be a contradiction. $\square$

4.7. The Complex Octagonal PETs

4.7.1. Complex Coordinates. We call the PETs $(\chi_1, \widehat{F})$ from §4.5 the *complex octagonal PETs* because it seems natural to interpret them as complexifications of the octagonal PETs. This statement is made more clear by the introduction of complex coordinates, as follows:

$$(x_1, x_2, x_3, x_4) = (x_1 + ix_3, x_2 + ix_4) = (z_1, z_2). \tag{4.39}$$

With these coordinates, we reinterpret the main formulas in §4.5.

Before we start, we remind the reader that a *real plane* in $\boldsymbol{C}^2$ is a 2-plane Π such that Π and $i\Pi$ are orthogonal. A *complex line* in $\boldsymbol{C}^2$ is a 2-plane Π such that $i\Pi$ and Π are parallel A *complex foliation* is a 2-dimensional foliation whose tangent planes are complex lines.

4.7.2. Basic Features. There are the basic features of the the complex octagonal PETs.

- The group G described above is generated by the reflections

$$\rho_1(z_1,z_2)=(z_1,-z_2), \qquad \rho_2(z_1,z_2)=(z_2,z_1). \tag{4.40}$$

  The invariant subspaces of ρ_1 and ρ_2 respectively are the complex lines given by $\{z_2=0\}$ and $\{z_1=z_2\}$.
- χ_1 is the parallelotope defined by the vectors

$$(2,0), \qquad (2i,0), \qquad (s\zeta,s\zeta), \qquad (s\overline{\zeta},s\overline{\zeta}); \qquad \zeta=1+i. \tag{4.41}$$

  The first two vectors lie in the complex line fixed by ρ_1 and the second two vectors lie in the complex line fixed by ρ_2.
- The foliation containing $\mathcal{R}_1$ is tangent to the basis $(\zeta,\overline{\zeta})$ and $(\overline{\zeta},-\zeta)$. The foliation containing $\mathcal{R}_2$ is tangent to the basis $(\overline{\zeta},\zeta)$ and $(\zeta,-\overline{\zeta})$. Both foliations are complex foliations.
- The involution ι is given by coordinate-wise conjugation. The invariant planes Π_0 and Π_2 for ι are the real planes $\boldsymbol{R}^2$ and $i\boldsymbol{R}^2$. Note that ι fixes Π_0 pointwise but acts as a reflection on Π_2. The restriction of $\widehat{F}$ to each of $\chi_1\cap\Pi_0$ and $\chi_1\cap\Pi_2$ is a copy of the octagonal PET at parameter s.

4.7.3. Additional Symmetry. The symmetry $\iota=\iota_1$ is an element of the order 8 dihedral group Γ generated by the two elements

$$\iota_1(z_1,z_2)=(\overline{z}_1,\overline{z}_2), \qquad \iota_2(z_1,z_2)=(iz_1,iz_2). \tag{4.42}$$

Each of these elements preserves $\chi_1,\chi_2,\Lambda_1,\Lambda_2$. Hence Γ acts as an order 8 groups of symmetries of the complex octagonal PET.

The existence of Γ in turn reveals more structure. There are 4 elements of Γ which act as real reflections – i.e., they pointwise fix real planes in $\boldsymbol{C}^2$. The planes $\Pi_0=\boldsymbol{R}^2$ and $\Pi_2=i\boldsymbol{R}^2$ are two of the fixed planes. The other two fixed planes are $\Pi_1=\zeta\boldsymbol{R}^2$ and $\Pi_3=\overline{\zeta}\boldsymbol{R}^2$. More simply,

$$\Pi_k=\zeta^k\boldsymbol{R}^2, \qquad k=0,1,2,3. \tag{4.43}$$

For instance, the map $\iota_2\circ\iota_1$ fixes Π_1 pointwise. The slices $\chi_1\cap\Pi_1$ and $\chi_1\cap\Pi_3$ are also invariant under $\widehat{F}$.

LEMMA 4.8. *There is a complex linear similarity S_s having the property that $S_s(\Pi_0)=\Pi_1)$ and $\widehat{F}_s^{-1}=S_s\circ\widehat{F}_{1/s}\circ S_s^{-1}$.*

Proof: We define

$$S_s=\frac{\zeta}{2s}\begin{bmatrix}1 & 1\\ 1 & -1\end{bmatrix}. \tag{4.44}$$

We compute that S_s has the following properties:

(1) S_s maps Π_k to Π_{k+1}, with indices taken mod 4.
(2) S_s maps $\chi_{j,s}$ to $\chi_{j,1/s}$ for $j=1,2$.
(3) S_s maps $\Lambda_{j,s}$ to $\Lambda_{3-j,1/s}$ for $j=1,2$.

Our result follows from these computations. □

In §7.5 we will prove that the octagonal PETs f_s and $f_{1/2s}$ are conjugate by a similarity. Combining this fact with the preceding result, and with symmetry, we get the following corollary.

COROLLARY 4.9. The systems $\widehat{F}_s|\Pi_1$ and $\widehat{F}_s|\Pi_3$ are conjugate, by similarities to the octagonal PET f_{2s}.

Thus, the complex octagonal PET $\widehat{F}_s$ contains 2 slices which are copies of f_s and two slices which are copies of f_{2s}.

CHAPTER 5

Outer Billiards on Semiregular Octagons

5.1. The Basic Sets

The Central Octagon: As in the introduction, let O_s be the semi-regular octagon having vertices

$$(5.1) \qquad (\pm s, \pm(1-s)), \qquad (\pm(1-s), \pm s).$$

Here $s \in (1/2, 1)$. We define the outer billiards map as in Figure 1.9. In this section we introduce some auxilliary sets.

The Barrier: Outer billiards relative to a polygon has a much simpler structure far away from the polygon. For the family O_s we can quantify this. We define *the barrier* to be the octagon having vertices

$$(5.2) \qquad (\pm 6, 0), \qquad (0, \pm 6), \qquad (\pm 6s, \pm 6s).$$

We mean for all possible sign choices to be made. Figure 5.1 shows a picture of the barrier.

The Fundamental Strips: For each edge e of O_s, we let Σ_s be the strip with the following characterization. One boundary line of Σ_e contains e. The other boundary line of Σ_s is such that the edge opposite e lies in the centerline of Σ. There are 8 such strips. Figure 5.1 below shows the strips and the barrier together.

The Far Domain: The barrier is chosen so that all the strip intersections take place in the barrier. 4 of these intersection points take place on the boundary of the barrier. Let e be the bottom edge of O_s. We define the *Far Domain* to be the region inside Σ_e and to the right of the barrier. Figure 5.1 shows the far domain.

Barrier Orbit: In Part IV, we will prove (Calculation 9) that there is a periodic tile whose orbit is a translate of O_s, such that the boundary of the barrier is contained in the union of the orbit. Figure 5.1 shows these tiles. In particular, the 8 vertices of the barrier are centers of some of these octagonal tiles. We emphasize that this structure holds for all $s \in (1/2, 1)$, though we are just showing the picture for $s = 3/4$.

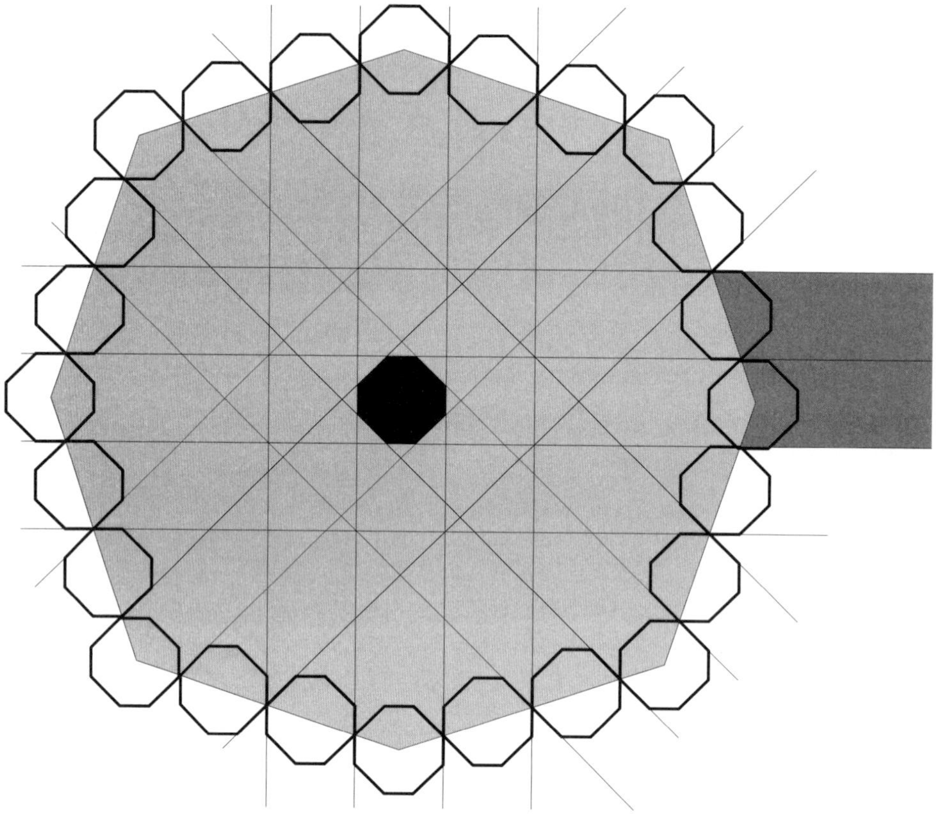

Figure 5.1: The fundamental objects for $s = 3/4$.

Here are the objects shown in Figure 5.1.

- O_s is black.
- The barrier is lightly shaded.
- The far domain darkly shaded.
- The strips are uncolored.
- The barrier orbit is uncolored.

Since the barrier orbit is invariant under the outer billiards map, and also covers the boundary of the barrier, we see that the orbit of any point outside the barrier orbit remains outside the barrier orbit forever.

5.2. The Far Partition

Now we will study the dynamics outside the barrier. We fix some parameter $s \in (1/2, 1)$ and then suppress it from our notation. Let ψ' be the outer billiards map and let $\psi = (\psi')^2$ be the second iterate. ψ is a piecewise translation. The partition for ψ consists of a finite union of compact polygons and convex noncompact polygonal sets.

LEMMA 5.1. *All the compact pieces of the domain for ψ are contained inside the barrier, and all the noncompact domains are bounded by segments or rays contained in the lines of the fundamental strips.*

Proof: Say that a *primary line* is a line extending a side of O_s. Say that a *secondary line* is the image of a primary line under reflection in one of the vertices of O_s. The union of secondary lines is precisely the set of lines in the fundamental strips described above.

Let $\mathcal{B}$ be the partition for ψ^2. Consider first the partition $\mathcal{A}$ of $\boldsymbol{R}^2$ into regions where both ψ' and $(\psi')^{-1}$ are completely defined. The regions of $\mathcal{A}$ are simply the complementary regions of the primary lines. The pieces of $\mathcal{B}$ are obtained by reflecting various pieces of $\mathcal{A}$ in various vertices of O_s. Hence, all the regions of $\mathcal{B}$ are bounded by segments or rays contained in the secondary lines.

The outermost intersections of secondary lines occur on the boundary of the barrier. Hence, none occur outside the barrier. This means that all the compact pieces of $\mathcal{B}$ are contained in the barrier. □

Let B denote the barrier. Let $F \subset \boldsymbol{R}^2 - B$ denote the far domain. The union of the boundaries of the fundamental strips divides $\boldsymbol{R}^2 - B$ into 32 noncompact pieces, which we call *the far partition* and denote by $\mathcal{F}$. Note that 24 of the pieces of $\mathcal{F}$ are half-strips, and the other 8 pieces are cones. F is a union of two of the half-strips of $\mathcal{F}$.

Each piece S of $\mathcal{F}$ is labeled by a vector V_S which has the property that $\psi(p) = p + V_S$ for all $p \in S$. Each half-strip of $\mathcal{F}$ intersects the barrier orbit. For this reason, we can deduce the labels of the pieces of $\mathcal{F}$ just by inspecting the barrier orbit. For instance, the upper half of the far domain gets the label $(0, -4s)$. The cone pieces of $\mathcal{F}$ are labeled according to the following rule: Each cone piece gets the same label as the half-strip which lies before it in the clockwise cyclic order.

5.3. The First Return Map

In this section we consider the dynamics outside the barrier. The dynamics is completly controlled by the 32 pieces of the far partition and their vector labels. We say that a *far orbit* is one which remains outside the barrier. From what we have said above, every point outside the barrier lies on a far orbit.

LEMMA 5.2. *Every far orbit intersects the far domain and the forward orbit of a point in the far domain returns to the far domain after one clockwise revolution around the barrier.*

Proof: Let B denote the barrier. We make 2 observations.

(1) For any point $p \in \boldsymbol{R}^2 - B$, the vectors $\psi(p) - p$ and $\overrightarrow{0p}$ form a positively oriented basls (same as $\{e_1, e_2\}$) and have a uniformly large angle between them. This is a precise way to say that, outside the barrier, the ψ-orbits circulate clockwise around O_s.
(2) Suppose $p \in \boldsymbol{R}^2 - B$ lies above the far domain. Then $\psi(p)$ cannot lie below the far domain. The vector labels do not have large enough y-components to skip over the far domain. Hence, ever orbit outside the barrier intersects the far domain every time it circulates around the barrier.

Our result follows immediately from these two observations. □

Let $\Psi = \psi|F$ be the first return map of ψ to the far domain F. Lemma 5.2 says that we can determine everything about the dynamics outside the barrier by looking at $\Psi : F \to F$.

As is familiar to people who have studied polygonal outer billiards on quasi-rational polygons, the map Ψ has a translational symmetry. This result follows from the *quasi-rationality* of the polygon. See [**VS**], [**K**], and [**GS**]. We will prove a precise result along these lines, from scratch.

Before we state and prove our lemma, we observe that the lower half of F and the 2 domains just below the lower half of F are both labeled by the vector

$$(4s-4, -4s). \tag{5.3}$$

These 3 pieces in a row all get the same label. This vector plays a crucial role in our arguments, and we will see it appear many times below.

LEMMA 5.3. *For any point $p \in F$, we have $\Psi(p+(0,4)) = \Psi(p)+(4,0)$.*

Proof: If p and q lie in the upper half of F, then

$$\psi(q) = q - (0,4s) = p + (4,0) - (0,4s) = \psi(p) + (4,0).$$

In this case, we replace p with $\psi(p)$. In this way we arrange that p and q lie in a region of $\mathcal{F}$ labeled by the vector $(4-4s, -4s)$, as discussed above.

For the first few iterates of ψ, we have the formula

$$\psi^{k+1}(q) - \psi^k(p) = (4s-4, -4s) + (4,0) = (4s, -4s). \tag{5.4}$$

This formula is valid until the two points on the left hand side of Equation 5.4 enter the upper fundamental strip of slope -1, as shown in Figure 5.2.

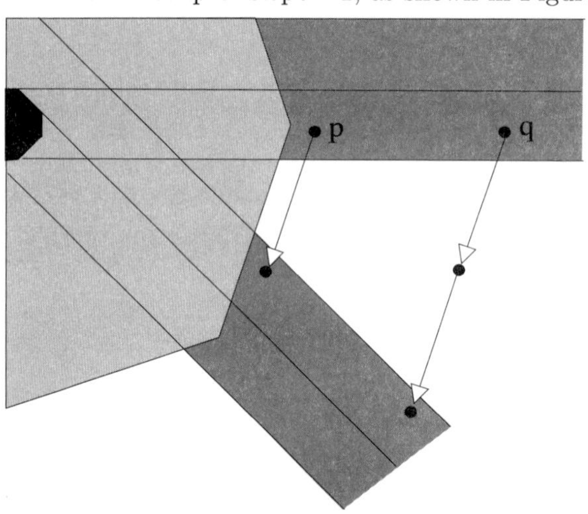

Figure 5.2: Return to the next strip.

The argument can now be repeated for points starting in the strip of slope -1.

We have gone 1/8 of the way around, so to speak. Applying a similar argument for the next leg of the journey, we see that

$$\psi^{m+2}(q) = \psi^m(p) + (0,-4) \tag{5.5}$$

the first time both orbits enter into the left vertical fundamental strip. Our lemma now follows from the 4-fold rotational symmetry of the picture. $\square$

Remark: In fact $\Psi(p+(2,0)) = \Psi(p)+(2,0)$ but this is more tedious to prove.

5.4. The Necklace Orbits

We say that a *necklace orbit* is a union $P_1, ..., P_n$ of octagons comprising the orbit of a single periodic tile, such that P_i and P_{i+1} share a vertex for all i, but otherwise the octagons are disjoint. Here the indices are taken cyclically. We insist that all the octagons are translation equivalent to O_s. The orbit shown in Figure 5.1 and guaranteed by Calculation 9, is a necklace orbit. Call this orbit Ω_6.

One of the octagons in Ω_6 has its center at the point $(6, 0)$. We check by direct calculation (Calculation 10) that there are necklace orbits Ω_n for $n = 2, 4, 6, 8, 10$. For each $n = 2, 4, 6, 8, 10$, there is an octagon of Ω_n centered at $(n, 0)$. See Figure 5.3 below. For $n = 8, 10$, these two octagons lie in the far domain.

LEMMA 5.4. *For each $n = 12, 14, 16, ...$ there exists a necklace orbit Ω_n, one of whose octagons is centered at $(n, 0)$.*

Proof: We will show that the existence of $\Omega(k)$ implies the existence of $\Omega(k+4)$ as long as $k \geq 8$ is congruent to 0 mod 4. For ease of exposition, we will show this for $k = 8$. The general case is the same. It follows immediately from Lemma 5.3 and the existence of Ω_8 that there is an orbit Ω_{12}, one of whose octagons is centered at $(12, 0)$ and is a translate of the central octagon. The argument given in the proof of Lemma 5.3 shows that the first three octagons of Ω_{12} are translates by $(4, 0)$ of the corresponding octagons in Ω_8. But then these three octagons touch vertex to vertex, as in a necklace. But then ψ maps the first octagon to the third one, and the second octagon to the fourth one. Since ψ just acts as a translation on these octagons, we see that the third and fourth octagon share a vertex. This pattern continues until two octagons in a row enter the upper fundamental strip of slope -1 (shown in in Figure 5.2.) Using the dihedral symmetry of the orbit, we see that Ω_{12} is a necklace. In short, the existence of Ω_8 implies the existence of Ω_{12}. $\square$

We have established the existence of necklace orbits $\Omega_2, \Omega_4, \Omega_6, ...$ There are $4n$ octagons in Ω_n. The regions between the necklace orbits, which we discuss below in detail, are invariant under the outer billiards map. In particular, all orbits are bounded. Compare the arguments in [**VS**], [**K**], and [**GS**].

5.5. Parallelograms, Halfbones, and Dogbones

Recall that Theorem 1.11 involves a tiling of $\boldsymbol{R}^2 - O_s$ by parallelograms which are translates of F_1, the domain of the octagonal PET at parameter s. The following result can be interpreted as saying that Theorem 1.11 is true for the central tiles of the octagonal PETs.

LEMMA 5.5. *Every octagon O' in a necklace orbit is contained in some parallelogram F' in the tiling from Theorem 1.11. The translation carrying F' to F_1 carries O' to the central tile $F_1 \cap F_2$ of the octagonal PET. Conversely, every parallelogram in the tiling contains an octagon from a necklace orbit.*

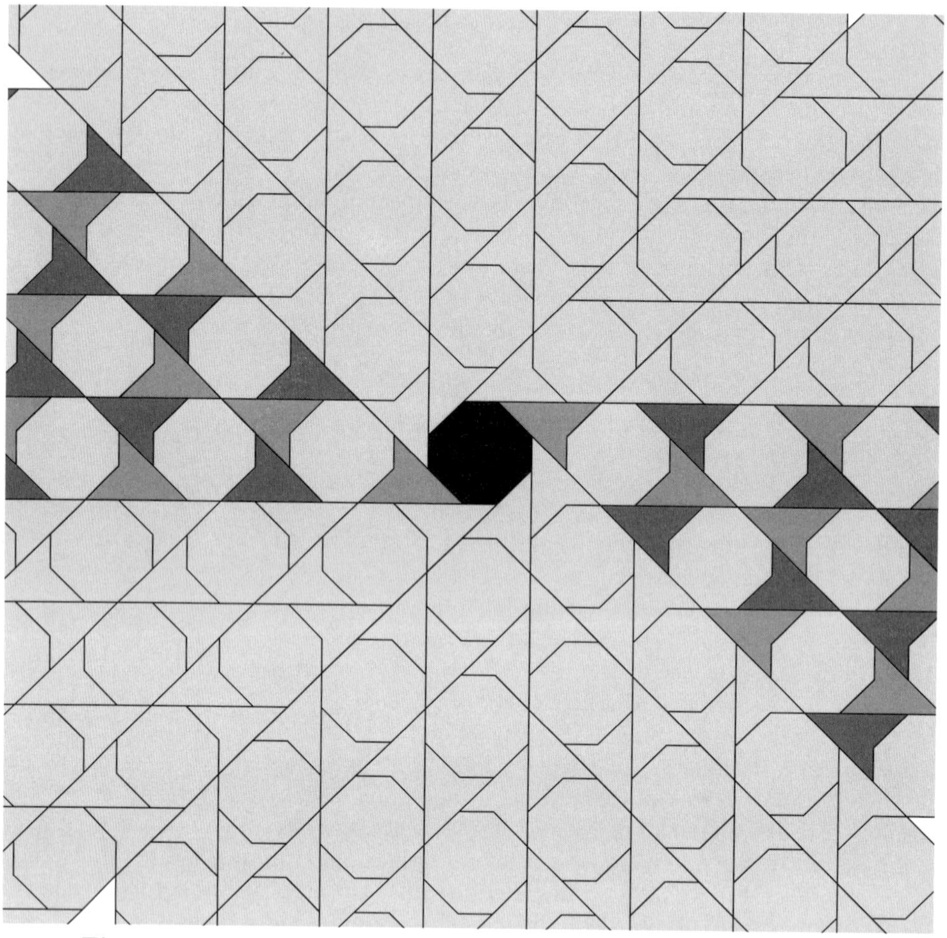

Figure 5.3: Parallelograms, Dogbones, and Halfbones for $s = 3/4$.

Proof: Our argument refers to figure 5.3. Let P_n be the octagon in Ω_n centered at $(n, 0)$. Our tiling is such that there is a parallelogram Π_n centered at $(n, 0)$ for $n = 2, 4, 6...$ (These parallelograms the top row of the right-hand shaded sector in Figure 5.3.) We check our result by hand for the pairs (P_n, Π_n) for $n = 2, 4$. But the map $p \to p + (2, 0)$ maps (P_n, Π_n) to (P_{n+1}, Π_{n+1}). So, by symmetry, our result holds for all octagons and parallelograms centered on the positive x-axis.

Our tiling is divided into 8 sectors, one of which is shaded in Figure 5.3. For the octagons in the shaded sector, the map

$$p \to p + (2 - 2s, -2s) \tag{5.6}$$

moves one octagon to the next and also moves one parallelogram to the next. The above symmetry now gives us our result for all the octagons and parallelograms in the shaded sector. Similar arguments work for the other sectors. □

The region Θ_{2n+1} between Ω_{2n} and Ω_{2n+2} is invariant under ψ. (This is the way one sees that all orbits are bounded.) Going outward from the central octagon, we encounter regions $\Theta_1, \Omega_2, \Theta_3, \Omega_4, ...$ The regions $\Theta_1, \Theta_3, \Theta_5$ lie in the barrier and the rest of them lie outside the barrier.

In view of Lemma 5.5, each region Θ_j is tiled by objects we call *halfbones*. A halfbone is a translate of a component of $F_1 \cap F_2$. Halfbones are nonconvex quadrilaterals. Figure 5.3 shows the picture for $s = 3/4$. We say that a *dogbone* is a union of two consecutive halfbones which share a common edge, provided that this union has bilateral symmetry. Each region Θ_n is partitioned into some finite union of dogbones and halfbones. (The partitions are not unique.)

We make the following observations.

- Each of the regions $\Theta_1, \Theta_3, \Theta_5, ..$ is partitioned into halfbones.
- Θ_1 contains no dogbones.
- No region Θ_n is *partitioned* into dogbones.
- The regions $\Theta_3, \Theta_5, \Theta_7$ are all unions of (overlapping) dogbones.

Whether a union of two adjacent halfbones is a dogbone is simply a combinatorial matter of how the two halfbones are oriented. Thus, the picture at a single parameter tells the whole story about which halfbones glue together to make dogbones. See Figure 5.7 below for another parameter.

5.6. The Dogbone Map

To the right of the central octagon, the top horizontal strip intersects Ω_n in a pair of dogbones, for $n = 3, 5, 7....$ Starting with $n = 5$, these dogbones lie in the far domain. Figure 5.4 shows the first few of these dogbones.

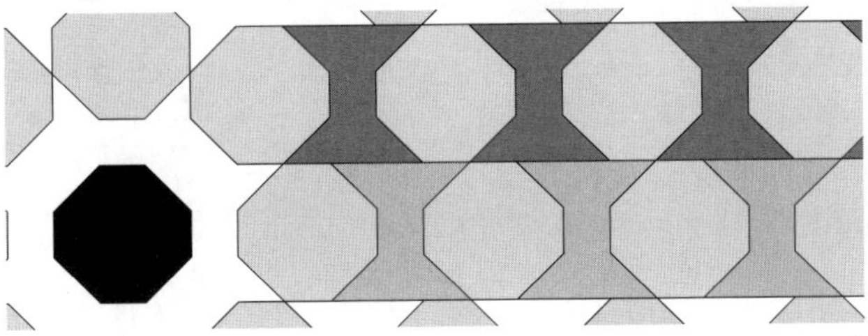

Figure 5.4: The row of dogbones for $s = 3/4$.

Let $D_n \subset \Omega_n$ denote the union of the two dogbones in the upper horizontal strip. We have

$$D_n = \Omega_n \cap F, \qquad n = 5, 7, 9... \tag{5.7}$$

Here F is the far domain. The case $n = 3$ is special, but what we say below applies to that case as well.

Recall that Ψ is the first return map of ψ (the second iterate of outer billiards) to F. Given that Ω_n is invariant under the outer billiards map, we have

$$\Psi : D_n \to D_n, \qquad n = 7, 9, 11... \tag{5.8}$$

Technically, we have not defined Ψ on D_n for $n = 3, 5$, but we do so now.

Remark: The arguments we gave about the structure of Ψ do not apply directly to the domains D_3 and D_5, because they lie inside the barrier. However, we can still write down the map $\Psi : D_3 \to D_3$ and $\Psi : D_5 \to D_5$ with the understanding that perhaps these maps are not entirely defined. That is, there might be some points which start in these domains and never return. In fact, our calculation will show that all points in D_3 and D_5 eventually do return to these regions. This calculation is part of Calculation 11, described below.

It is tempting to show that the first return map to D_n is conjugate to the octagonal PET at the corresponding parameter. However, this does not work directly. We need to massage the picture a bit to make things work.

Looking at Figure 5.4, we can see that it makes sense to write

$$D_n = D_n^0 \cup D_n^1 \tag{5.9}$$

where D_n^0 is the lower dogbone and D_n^1 is the upper dogbone.

There is an obvious involution $\iota : D_n \to D_n$ which simply interchanges the two dogbones using a piecewise translation. As part of Calculation 11, described below, we will show that

$$\Psi \circ \iota = \iota \circ \Psi \tag{5.10}$$

on D_n for $n = 3, 5, 7, 9$. Equation 5.10 then holds for $n = 11, 13, 15, ...$ by Lemma 5.3. Equation 5.10 allows us to "compress" the map Ψ so that it really just lives on a single dogbone.

We define a new map $\Upsilon : D_n^0 \to D_n^0$ by the rule

- $\Upsilon(p) = \Psi(p)$ if $\Psi(p) \in D_n^0$.
- $\Upsilon(p) = \iota \circ \Psi(p)$ if $\Psi(p) \in D_n^0$.

We call $\Upsilon(p)$ the dogbone map.

LEMMA 5.6. *The periodic tiling of D_n^0 relative to Υ is just the restriction of the outer billiards periodic tiling to D_n^0.*

Proof: It follows immediately from Equation 5.10 that the Υ-orbit of p is periodic (respectively aperiodic or undefined) if and only if the Ψ-orbit of p is periodic (respectively aperiodic or undefined). Hence, the periodic tiling of D_n^0 relative to Υ is the same as the periodic tiling relative to Ψ. But the periodic tiling with respect to Ψ is the same as the periodic tiling with respect ψ, because Ψ is just the first return map of ψ to some domain. Finally, the periodic tiling with respect to ψ is the same as the periodic tiling with respect to ψ' because ψ is the second iterate of ψ'. $\square$

5.7. The First Conjugacy

As usual, all our constructions depend on a parameter $s \in (1/2, 1)$, which we sometimes suppress from our notation.

Let F_1 and F_2 be the parallelograms associated to the octagonal PET. We have a partition

$$F_1 = H_1 \cup (F_1 \cap F_2) \cup H_2 \tag{5.11}$$

where $F_1 \cap F_2 = O_s$ is isometric to the octagons we have been considering above and each of H_1 and H_2 is isometric to a halfbone. We define

$$D = H_2 \cup (H_1 + (2, 0)). \tag{5.12}$$

D is the dogbone shown on the right hand side of Figure 5.5.

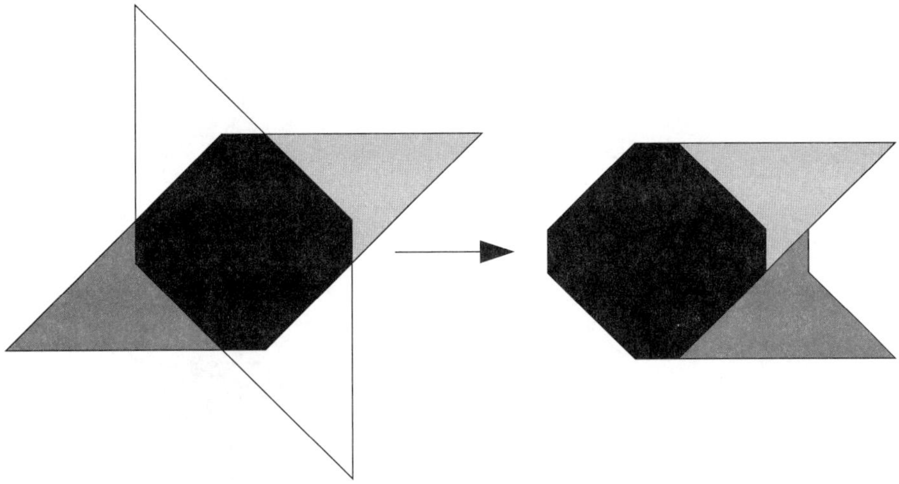

Figure 5.5: H_1 (dark) and $F_1 \cap F_2$ (black) and H_2 (light).

Let f be the octagonal PET. Since $H_1 \cup H_2$ is an invariant set for f, we can take D as the domain for $f|(H_1 \cup H_2)$. More precisely, there is a piecewise isometry from $H_1 \cup H_2$ to D, and we consider the map on D which is conjugate to f by this piecewise isometry. Conveniently, the map

$$\theta_n(p) = p + (0, n) \tag{5.13}$$

maps D to D_n^0 for $n = 0, 2, 4, \ldots$. Here D_n^0 is the bottom half of the domain D_n.

In Part IV we establish the following result by direct calculation.

LEMMA 5.7 (Calculation 11). *The following is true for $n = 3, 5, 7, 9$:*

- *Equation* 5.10 *holds on D_n.*
- *θ_n conjugates f^2 to Υ on D_n.*

In particular, Υ is well defined on almost all points of D_n.

COROLLARY 5.8. The periodic tiling on D relative to f is isometric to the restriction of the outer billiards periodic tiling to D_n for all $n = 3, 5, 7 \ldots$.

Proof: Combining Lemma 5.7 with Lemma 5.3, we see that in fact Lemma 5.7 is true for all $n = 3, 5, 7, \ldots$. Hence the tiling on D relative to f^2 is isometric to the tiling on D_n^0 relative to Υ (and relative to outer billiards). Finally, the second iterate f^2 produces the same tiling on D as f does. □

Before we go further, we remark on an extra symmetry of the outer billiards tiling which would not be so easy to see without Corollary 5.8. At the same time, we get an extra symmetry of the octagonal PET which we would not notice (or, rather, know how to prove) without the connection to outer billiards.

COROLLARY 5.9. For $n = 3, 5, 7, \ldots$ the restriction of the outer billiards tiling to D_n^0. has 4-fold dihedral symmetry.

Proof: Reflection in the origin commutes with all the objects defining the octagonal PET. Hence the tiling of F_1 relative to f has rotational symmetry. Hence the tiling of D relative to f has rotational symmetry. Hence the tiling of D_n^0 relative to outer billiards has rotational symmetry.

At the same time, reflection in the x-axis conjugates the outer billiards map to its inverse. Hence, the tiling of D_n^0 relative to outer billiards has reflection symmetry. The reflection and rotation generate an order 4 dijedral group of symmetries. □

Figure 1.2 shows the picture for a rational parameter very close to the parameter $u = 3/2 - \sqrt{3}/2$. Note that $1 - u$ is oddly even, so (by the rational version of Statement 2 of Corollary 1.3) all the tiles in Figure 5.6 except the very smallest ones are squares.

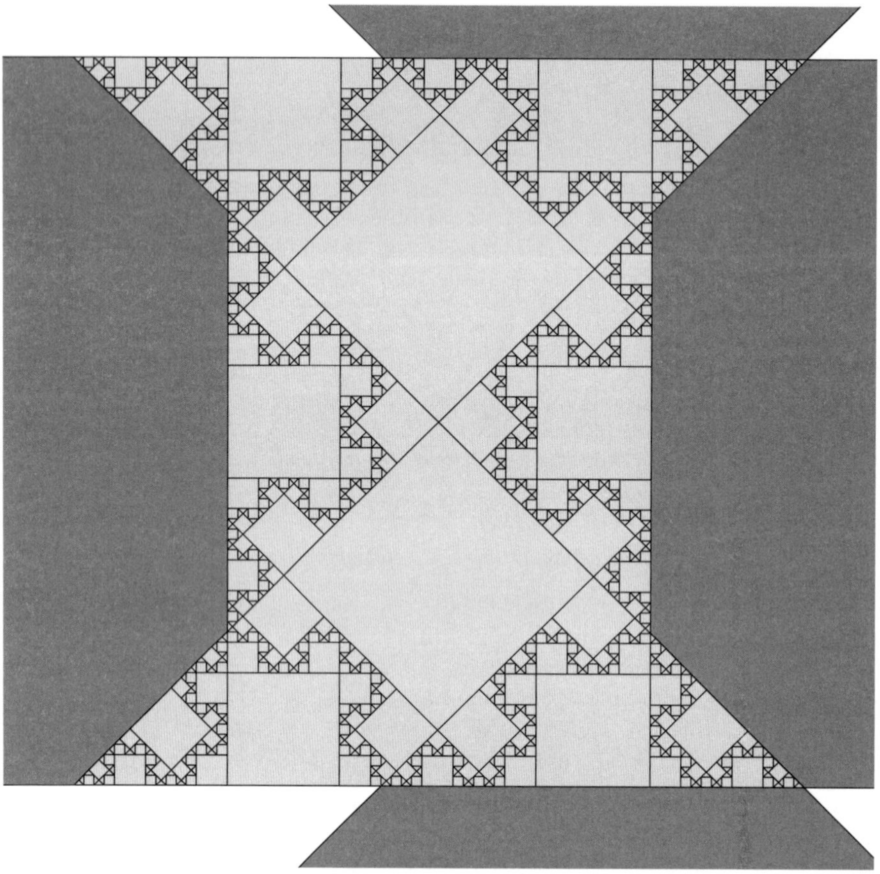

Figure 5.6: The tiling of D_3^0 for $s = 71/112$.

Say that a parallelogram is *clean* if it does not intersect Ω_1. All parallelograms but the innermost 8 are clean. Let Σ be the horizontal strip associated to O_s which goes off to the right. Σ contains the dogbones shown in Figure 5.4. Note that Σ contains two rows of dogbones, a lower row and an upper row. All the parallelograms in Σ are clean except for the one closest to O_s. For convenience we repeat Figure 5.4.

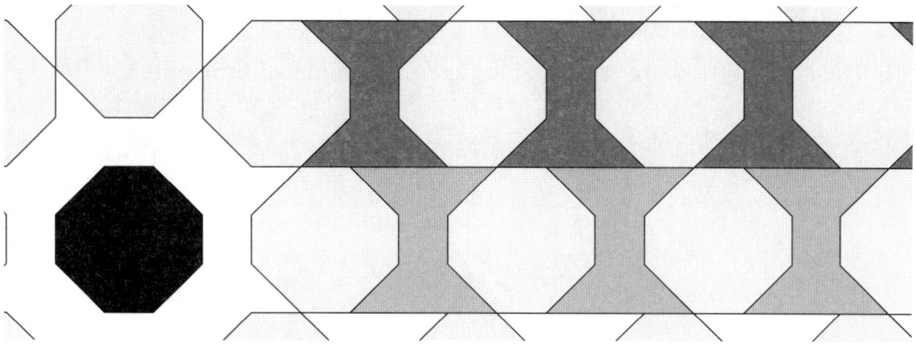

Figure 5.4: The row of dogbones for $s = 3/4$.

LEMMA 5.10. *Theorem* 1.11 *holds for all the clean parallelograms in* Σ.

Proof: Recall that Π_n is the parallelogram in our tiling which contains the octagon of Ω_n that is centered at $(n, 0)$. Note that Π_n is clean for all values $n = 4, 6, 8, \ldots$ but not for $n = 2$. Let us consider Π_4. We have the decomposition

$$\Pi_4 = H_{3U} \cup P_4 \cup H_{5L}. \tag{5.14}$$

Here H_{3U} is the upper halfbone of D_3^0 and H_{5L} is the lower halfbone of D_5^0. It follows from what we have already proven, and symmetry, that the tiling of F_1 relative to the octagonal PET is isometric to the tiling of Π_4 relative to outer billards. But the same argument works for Π_n for $n = 6, 8, 10, \ldots$. Now we know Theorem 1.11 for all the clean parallelograms in the lower half of Σ.

Given Equation 5.10, everything we said about the outer billiards tiling in D_n^0 for $n = 3, 5, 7, \ldots$ holds inside D_n^1 for $n = 3, 5, 7, \ldots$. The same cut-and-paste argument as above shows that Theorem 1.11 holds for all the clean parallelograms in the upper half of Σ. Hence Theorem 1.11 holds for all the clean parallelograms in Σ. □

LEMMA 5.11. *Theorem* 1.11 *holds for all the clean parallelograms.*

Proof: By symmetry, Theorem 1.11 holds for all the clean parallelograms in all 8 strips defined by O_s. The argument here is a bit subtle. The tiling in the diagonal strips associated to O_s is isometric to the tiling in the horizontal and vertical strips associated to O_t, where $t = 1/(2s)$. This subtlety does not bother us, because Lemma 5.10 holds for all $s \in (1/2, 1)$.

Let Z be an arbitrary clean parallelogram. In view of our description of the outer billiards dynamics, there is a clean parallelogram Z', contained in one of the strips, such that the tiling in Z is just a translate of the tiling in Z'. Hence, Theorem 1.11 holds for Z. □

It only remains to prove Theorem 1.11 for the innermost 8 parallelograms. We will establish this for the parallelogram Π_2. Once we know the result for Π_2 we get the remaining 7 by symmetry. That will finish the proof of Theorem 1.11.

5.8. The Second Conjugacy

To deal with Π_2 we have to understand what happens in the domain Θ_1. This domain contains no dogbones, so it is hard to find a direct conjugacy to the octagonal PET. Instead we proceed as follows. Let $E_1 \subset \Omega$ be the shaded region shown in Figure 5.7. It appears that E_1 is the union of a dogbone and two halfbones, but the dark central piece is not a union of halfbones which come from the parallelogram tiling. Even so, let $D_3 = E_1 + (2, 0)$.

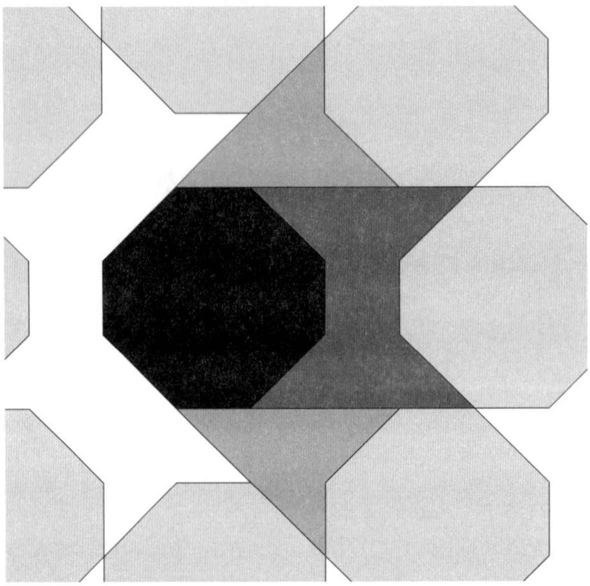

Figure 5.7: Close-up of Θ_s for $s = 4/5$.

In Part IV we prove the following result by direct calculation.

LEMMA 5.12 (Calculation 12). *The map $p \to p + (2, 0)$ conjugates $\Psi|E_1$ to $\Psi|E_3$.*

It follows from Lemma 5.12 that the restriction of the outer billiards tiling to the halfbones of E_1 is isometric to the restriction of the outer billiards tiling to the halfbones of E_3. In particular, this gives us Theorem 1.11 for the prallelogram Π_2. This completes the proof of Theorem 1.11.

CHAPTER 6

Quarter Turn Compositions

In this chapter, we generalize some of the ideas discussed in previous chapters. The results here are proved in full in [**S2**]. Our main purpose is to illustrate the close connection between polygonal outer billiards and double lattice PETs. The common theme is provided by a family of maps which we call *quarter turn compositions*. These are similar to the alternating grid systems.

6.1. Basic Definitions

We will work in the infinite strip

$$\mathbf{S} = \boldsymbol{R} \times [-1/2, 1/2]. \tag{6.1}$$

We are going to define two kinds of maps of $\mathbf{S}$, shears and quarter turns.

We define the *shear*

$$S_s = \begin{bmatrix} 1 & -s \\ 0 & 1 \end{bmatrix}, \qquad s > 0. \tag{6.2}$$

The map S_s is a shear of $\mathbf{S}$ which fixes the centerline pointwise, moves points with positive y-coordinate backwards and points with negative y-coordinate forwards.

Let $\square$ be a rectangle with sides parallel to the coordinate axes. We define a *quarter turn* of $\square$ to be the order 4 affine automorphism of $\square$ which maps the right edge of $\square$ to the bottom edge of $\square$. For any $a > 0$ we distinguish 2 tilings of the strip $\mathbf{S}$ by $a \times 1$ rectangles. In *Tiling 0*, the origin is the center of a rectangle. In *Tiling 1*, the origin is the center of a vertical edge of a rectangle. For $q = 0, 1$ let $R_{q,a}$ denote the map which gives a quarter turn to each rectangle in Tiling q. We call the piecewise affine map $R_{q,a}$ a *quarter turn*.

We define a *quarter turn composition* (QTC) to be a finite alternating composition $\mathcal{T}$ of quarter turns and shears. That is,

$$\mathcal{T} = S_{s_n} \circ R_{q_n, r_n} \circ \cdots \circ S_{s_1} \circ R_{q_1, r_1}. \tag{6.3}$$

- $q_1, ..., q_n \in \{0, 1\}$ specify the tiling offsets.
- $r_1, ..., r_n$ are the parameters for the widths of the rectangles.
- $s_1, ..., s_n > 0$ are the parameters for the shears.

We call n the *length* of the QTC.

6.2. The Polytope Graph

Here we explain a generalization of the concept of a multigraph PET. The construction suggests a definition of 3-dimensional polyhedral outer billiards, at least for fairly generic polyhedra.

Partial Lattices and Extended Polytopes: Say that a *partial lattice* is an abelian subgroup of rank $k < n$ contained in $\boldsymbol{R}^n$. (For $k = 1$, we are talking about $\boldsymbol{Z}(V)$ for some vector V.) Say that an *extended polytope* is a subset of $\boldsymbol{R}^n$ which is isometric to the product $P^k \times \boldsymbol{R}^{n-k}$ where P^k is a k-dimensional convex polytope. (For $k = 1$ we get an infinite strip.)

We say that an extended polytope is a *fundamental domain* a partial lattice if the translates of the extended polytope by the vectors in the partial lattice forms a tiling of $\boldsymbol{R}^n$. (For $k = 1$ we require that the vector generating the lattice can be placed so that the head lies on one edge of the strip and the tail lies on the other.)

Generalized Decorations: We generalize the construction of a decorated multigraph by allowing the vertices to be labelled by extended polytopes and the edges to be labelled by partial lattices. We require the basic property that a vertex is incident to an edge if and only if the corresponding extended polytope is a fundamenta domain for the corresponding partial lattice. With this generalization, we get the same functors as above, provided we enlarge the category **PET** so that it includes maps on partial polytopes which are piecewise translations.

The Facet Graph: Let Ω be a convex polytope. By a *facet* of a polytope we mean a codimension 1 face. Suppose, for each facet f there is a unique vertex $v(f)$ of Ω which is farthest from the hyperplane extending f. We form a graph $\Gamma = \Gamma(\Omega)$ as follows. The vertices of Γ are the facets of Ω. We join two vertices f_1 and f_2 of Γ if and only if $v(f_1)$ is a vertex of f_2 and $v(f_2)$ is a vertex of f_1. (When Ω is a polygon with no parallel sides, the facet graph is a cycle.)

We decorate Γ as follows. For each vertex f we assign the partial polytope P_f which is an infinite slab bounded by parallel hyperplanes. One of the hyperplanes is the extension of the face f. The other hyperplane is parallel to the first one, so that $v(f)$ is equidistant between the two hyperplanes. To the edge joining f_1 and f_2 we assign the lattice

$$\boldsymbol{Z}(V), \qquad V = 2(v(f_1) - v(f_2)). \tag{6.4}$$

Note that this definition is independent of the order in which we choose f_1 and f_2.

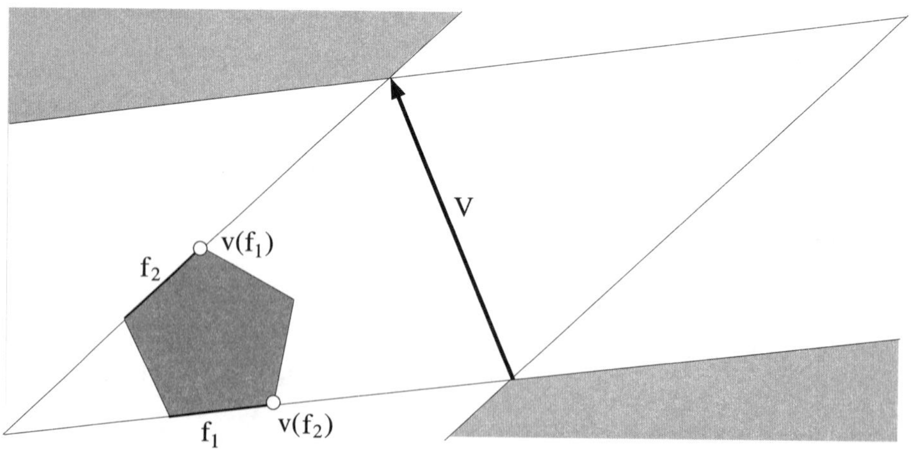

Figure 6.1: Constructing the facet graph

Figure 6.1 shows the example of the regular pentagon. We show facets f_1 and f_2, and the corresponding strips, and the vector V. The two strips are the extended polygons associated to f_1 and f_2. Notice that both strips are fundamental domains for $\boldsymbol{Z}(V)$. The picture looks similar for any polygon without parallel sides.

The Tetrahedron: Consider the case of a tetrahedron τ. In this case, the facet graph Γ is the complete graph on 4 vertices. Each pair of facets (f_1, f_2) of τ meet in an edge, and the opposite edge, when suitably oriented, equals the vector $v(f_1) - v(f_2)$. Once we choose a distinguished facet of Γ, corresponding to a vertex γ of Γ, and suitably scale the picture, we get an action of the fundamental group $\pi_1(\Gamma, \gamma)$ on the infinite slab $\boldsymbol{R}^2 \times [0,1]$. Here $\pi_1(\Gamma, \gamma)$ is isomorphic to the free group on 3-generators. This seems like an appealing example to study. For reasons which will emerge below, we would call this system "outer billiards on a tetrahedron".

6.3. QTCs and Polygon Graphs

In the 2-dimensional version of our construction, Ω is an n-gon with no parallel sides, Γ is an n-cycle, and the vertices of Γ are labelled by infinite strips $\Sigma_1, ...\Sigma_n$. The corresponding faces are $f_1, ..., f_n$. The edges are labeled by $\boldsymbol{Z}(V_k)$, where $V_k = 2v(f_k) - 2v(f_{k-1})$.

We fix one of the strips of Γ, say $\Sigma = \Sigma_n$. Let f be the corresponding vertex of Γ. We get a (generalized) PET $\xi : \Sigma \to \Sigma$. Here ξ is the image of 1 under the composition $\boldsymbol{Z} \leftrightarrow \pi_1(\Gamma, f) \to \mathrm{PET}(\Sigma)$. (This defines ξ up to taking inverses.) In [**S2**] we prove the following result.

LEMMA 6.1. *There is an affine transformation from Σ to* **S** *which conjugates ξ^2 to a the map $\mathcal{T}^2$, where $\mathcal{T}$ is a quarter turn composition.*

Proof: (sketch) Consider the elementary maps

$$T_k = \Sigma_{k-1} \to_{V_k} \Sigma_k. \tag{6.5}$$

In order to consider the second iterate, we define $\Sigma_{k+n} = \Sigma_k$, etc.

We have

$$\xi^2 = T_{2n} \circ T_{2n-1} \circ \cdots \circ T_1. \tag{6.6}$$

For $k = 1, ..., 2n$ we define area- and orientation-preserving affine maps

$$A_{k,\pm} : \Sigma_k \to \mathbf{S}, \tag{6.7}$$

such that $A_{k,\pm}$ carries $\Sigma_{k\pm 1} \cap \Sigma$ to a rectangle and $v(f_k)$ to the origin.

Let $\rho(x, y) = (-x, -y)$ be reflection about the origin. Our description determines $A_{k,\pm}$ up to composition with ρ. It turns out that one can make consistent choices so that the two maps

$$R_k = A_{k+1,-} \circ T_k \circ A_{k,+}^{-1}; \qquad S_k = A_{k+1,+} \circ (A_{k+1,-})^{-1} \tag{6.8}$$

respectively are affine shears and quarter turn maps. We need to go around twice to make the choices consistent because $A_{n+k,\pm} = \rho \circ A_{k,\pm}$.

Let

$$\mathcal{T} = S_n \circ R_n \circ \cdots \circ S_1 \circ R_1. \tag{6.9}$$

When we compose these maps there is a lot of cancellation, and we get $\mathcal{T}^2 = S_{2n}\xi^2 S_0^{-1} = S_0\xi^2 S_0^{-1}$. □

We say that $\mathcal{T}$ is *finitary* if the set of differences

$$\{\mathcal{T}(p) - p| \ p \in \mathbf{S}\} \tag{6.10}$$

is a finite set.

LEMMA 6.2. *$\mathcal{T}^2$ is finitary.*

Proof: Our whole construction is affinely invariant, so we normalize P so that in fact $\xi^2 = \mathcal{T}^2$. We use the notation from the proof of Lemma 6.1. Let $p = p_0 \in \Sigma_0$ and define

$$p_k = T_k \circ \ldots \circ T_1(p_0). \tag{6.11}$$

Let m_k be such that

$$p_k = p_{k-1} + m_k V_k. \tag{6.12}$$

Then

$$\mathcal{T}^2(p) - p = \sum_{i=1}^{2n} m_i V_i = \sum_{i=1}^{n} (m_i + m_{i+n}) V_i. \tag{6.13}$$

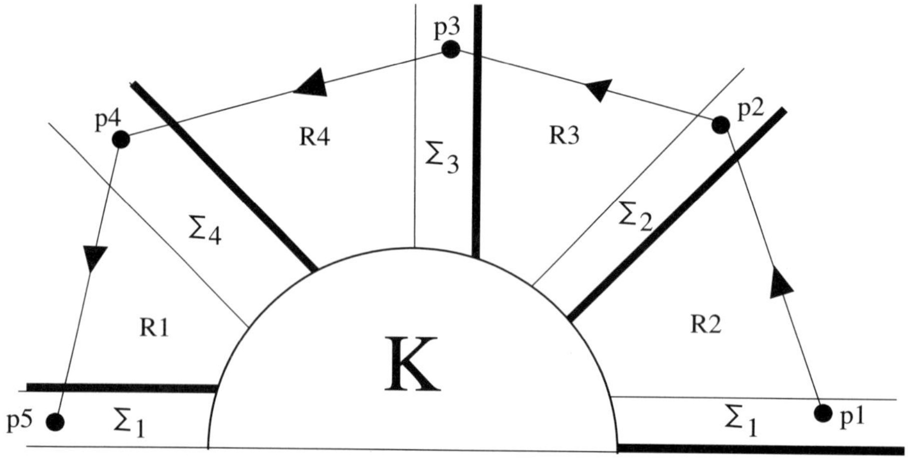

Figure 6.2: Far from the origin.

Outside of a large compact set K, the portion of the ξ-orbit of p, going from p to $T^2(p)$, lies within a uniformly bounded distance of a centrally symmetric $2n$-gon. The point is that all the strips come within a uniform distance of the origin. This means that there is a uniform bound to $|m_i + m_{i+n}|$ for all i. Hence, there are only finitely many choices for $\mathcal{T}^2(p) - p$. □

6.4. QTCs and Outer Billiards

Let Ω be a convex polygon with no parallel sides, as in the previous section. On the outer billiards side, the constructions given in §5.1 generalize (to some extent) to any convex polygon. However, on the QTC side, we have only worked out the picture for polygons having no parallel sides.

In general, there may not be a "barrier" which separates "far orbits" from "near orbits". For instance, in [**S3**] we proved the following result.

THEOREM 6.3 (Erratic Orbits). *Let K be any irrational kite. Then there exists an unbounded outer billiards orbit relative to K which has a vertex of K as a subsequential limit.*

A kite is irrational if one of its diagonals divides it into regions having irrationally related areas. The orbits guaranteed by the Erratic Orbits Theorem will pass through any barrier one might try to place around the kite.

On the other hand, there is always a barrier in the weaker sense that the first return map to one of the half-strips (the *far domain* from §5.1) has a quite simple description: The orbit just circulates once around Ω and comes back to the strip in roughly the same place. However, after many revolutions around, the orbit can drift towards Ω, as in the Erratic Orbits theorem.

Let ψ' denote the outer billiards map relative to Ω and let $\psi = (\psi')^2$. The map ψ is always a translate by a vector of the form $2v - 2w$, where v and w are vertices of Ω. Far from the origin, f^2 is a translate by some $\pm V_k$, where $V_1, ..., V_n$ are the vectors considered in the previous section.

Indeed, let Σ be the strip considered in the previous section. Let Ψ' denote the first return map of ψ to Σ. (Here we are deliberately considering the first return map of ψ, and not ψ'.) Let $\Psi = (\Psi')^2$. So, Ψ is the second return of the second iterate of the outer billiards map.

The two maps Ψ and ξ^2 coincide outside a large compact set $K = K_\Omega$. Here K plays the role of the barrier from §5.1. What we are saying is that Figure 6.2 describes both maps ξ^2 and Ψ. Our result [**S4**, Pinwheel Theorem] gives much tighter control on the relation between ξ^2 and Ψ.

THEOREM 6.4 (Pinwheel). *There is a canonical bijection between the set of unbounded orbits of ξ^2 and the set of unbounded orbits of Ψ. The bijection is such that corresponding orbits agree outside a compact set. In particular ξ has unbounded orbits if and only if Ψ does.*

The point of the Pinwheel Theorem is that, up to some bounded error, every polygonal outer billiards map can be expressed in terms of a quarter turn composition. As we already mentioned above, part of the Pinwheel Theorem is easy: outside of K, the maps ξ^2 and Ψ coincide.

The analysis of what happens fairly near Ω is intricate, and it seems miraculous that it works out: The ξ^2 orbit and the Ψ orbit of a point can look quite different when the point is near Ω. We think of outer billiards near Ω as being like the complicated part of a pinball machine. Points wander into the complicated part and get banged around, but then they emerge later on in a way that is predicted by the much simpler QTC.

It is interesting to compare the general case to what happens for semi-regular octagons. In the case of the semi-regular octagons, there is an infinite sequence of "necklace orbits" which confine the remaining orbits between them. This happens for any quasi-rational polygon, according to the results in [**VS**], [**K**], and [**GS**]. The polygon Ω is *quasi-rational* if one can scale Ω so that the parallelograms $\Sigma_k \cap \Sigma_{k+1}$ have integer area for all k. In general there are no such necklace orbits, and the Erratic Orbits Theorem shows that in fact there can be unbounded orbits.

Given the connection between QTCs and both outer billiards and the PETs which arise from the polytope graph in 2 dimensions, we see that there is a connection between outer billiards and the PETs which arise from the polytope graph. This suggests that one way to *define* outer billiards in higher dimensions is to start with a polytope and consider the subcategory of PETs defined by the associated polytope graph.

One shortcoming of our definition is that it does not work so gracefully when the polytope has parallel faces (of any dimension). This is disappointing, because one would like to try things out for the platonic solids, and 4 out of 5 of them have parallel facets. Probably there is a way to extend our definition to the case where there are some parallel faces, but we have not tried to do it.

There is one more thing we would like to say in connection with higher dimensional outer billiards. Our proposal for a higher dimensional definition is perhaps only useful in odd dimensions. Sergei Tabachnikov has pointed out that one can use the complex structure on $\boldsymbol{R}^{2n} = \boldsymbol{C}^n$ to make a well-defined outer billiards map in all even dimensions. These higher dimensional systems, both his and ours, are completely unexplored.

6.5. QTCs and Double Lattice PETs

QTCs are similar in spirit to the alternating grid systems considered in the previous chapter. Here they have the added complication that the maps involved are affine maps rather than isometries. However, if $\mathcal{T}$ is a quarter turn composition corresponding to a polygonal outer billiards system, then $\mathcal{T}^2$ is a finitary piecewise translation. One would expect that such a map has a compactification which is a higher dimensional PET. In **[S2]** we prove the following result.

THEOREM 6.5 (Compactification). *Suppose that $\mathcal{T}$ is a length-n quarter turn composition with parameters $\{a_i\}$ and $\{r_i\}$ and $\{s_i\}$. Suppose also that $\mathcal{T}^2$ is a finitary piecewise isometry. Then $(\mathbf{S}, \mathcal{T})$ has a compactification which is a double lattice PET. The dimension of the compactification is the dimension over $\boldsymbol{Q}$ of the vector space $\boldsymbol{Q}(r_1/r_n, ..., r_{n-1}/r_n)$.*

Proof: (sketch) The proof of Theorem 6.5 is very similar to the proof of Theorem 1.12 given in the previous chapter. Let $\widehat{\mathbf{S}} = \boldsymbol{R}^{n+1}/\boldsymbol{Z}^{n+1}$.

$$\Psi(x, y) = \left(\frac{x}{r_1}, ..., \frac{x}{r_n}, y\right) \bmod \boldsymbol{Z}^{n+1}. \tag{6.14}$$

Referring to Equation 6.9, the first step in our proof here is to show that there are affine maps $\widehat{R}_1, ..., \widehat{R}_n, \widehat{S}_1, ..., \widehat{S}_n$ of $\widehat{\mathbf{S}}$ such that

$$\Psi \circ R_k = \widehat{R}_k \circ \Psi, \qquad \Psi \circ S_k = \widehat{S}_k \circ \Psi. \tag{6.15}$$

This is done, as in the previous chapter, by finding explicit formulas for these map. These maps depend on the parameters $a_1, ..., a_n$ and $s_1, ..., s_n$, but only in a mild way. For different parameters, we get the same maps up to composition with a translation of $\widehat{\mathcal{S}}$.

Once we have the maps $\widehat{R}_k$ and $\widehat{S}_k$, for $k = 1, ..., n$, we define $\widehat{\mathcal{T}}$ exactly as we defined $\mathcal{T}$, but with respect to these new maps. Below we will explain how we use the finitary condition to prove that $(\widehat{\mathcal{T}}^2, \widehat{\mathbf{S}})$ is a piecewise translation. An analysis of the singularity set of $\widehat{\mathcal{T}}$, along the lines of Lemma 4.3, allows us to replace the domain $\boldsymbol{T}$ by a parallelotope and recognize $\widehat{\mathcal{T}}$ as a double lattice PET. □

Using the Finitary Property: It seems worthwhile explaining how the finitary condition implies that $\widehat{\mathcal{T}}^2$ is a local translation. For simplicity, we will treat the (generic) case when $\Psi(\mathbf{S})$ is dense in $\widehat{\mathbf{S}}$. In this case, it suffices to prove that the restriction of $\mathcal{T}^2$ to $\Psi(\mathbf{S})$ is a local translation.

Suppose that $\widehat{p} \in \Psi(\mathbf{S})$ and $\{\widehat{p}_n\} \in \Psi(\mathcal{S})$ is a sequence of points converging to $\widehat{p}$. We want to show that

$$\widehat{\mathcal{T}}(\widehat{p}) - \widehat{p} = \widehat{\mathcal{T}}(\widehat{p}_n) - \widehat{p}_n, \tag{6.16}$$

for all n sufficiently large. We have points $p, p_n \in \mathbf{S}$ such that $\widehat{p} = \Psi(p)$ and $\widehat{p}_n = \Psi(p_n)$. Note that $\{p_n\}$ need not be a convergent sequence in $\mathbf{S}$.

LEMMA 6.6. *Let $V_q = \mathcal{T}(q) - q$ for any $q \in S$, we have*

$$\widehat{\mathcal{T}}(\widehat{p}) - \widehat{p} = \Psi(V_p), \qquad \widehat{\mathcal{T}}(\widehat{p}_n) - \widehat{p}_n = \Psi(V_{p_n}). \tag{6.17}$$

Proof: We have

$$\widehat{\mathcal{T}}(\widehat{p}) - \widehat{p} = \widehat{\mathcal{T}} \circ \Psi(p) - \Psi(p) = \Psi \circ \mathcal{T}(p) - \Psi(p) = \Psi(V_p). \tag{6.18}$$

The last equality comes from the fact that $\Psi(V + W) = \Psi(V) + \Psi(W)$ whenever V, W, and $V + W$ all belong to $\mathbf{S}$. Here we are taking $V = V_p$ and $W = p$. The same argument works for p_n. □

We now observe the following properties:

(1) By continuity and Equation 6.17, we have $\Psi(V_{p_n}) \to \Psi(V_p)$ as $n \to \infty$.
(2) Since $\mathcal{T}$ is finitary, there is a uniform upper bound to $|V_{p_n}|$.
(3) Ψ is injective.

It follows from these properties that $V_{p_n} \to V_p$. But $\mathcal{T}$ is finitary. Hence $V_{p_n} = V_p$ for n large. But then $\Psi(V_p) = \Psi(V_{p_n})$ for n large. This fact combines with Equation 6.17 to establish Equation 6.16 for n large.

Summary: The first return map to the strip Σ captures all the essential information about the dynamics of a polygonal outer billiards system. The first return map just mentioned is equivalent, in a strong sense, to a quarter turn composition. The quarter turn composition in turn has a compactification which is a double lattice PET. Thus, many features of a polygonal outer billiards system, including the existence of unbounded orbits, can be studied by looking at a suitable double lattice PET.

Part 2

Renormalization and Symmetry

The main goal of this part of the monograph is to introduce the elementary symmetry properties of the octagonal PET and to prove the Main Theorem.

- In §7 we explain some elementary symmetry properties enjoyed by the octagonal PETs.
- In §8 we introduce a combinatorical gadget which we call the *arithmetic graph*. We use this gadget to show that the tiles in Δ_s, for any irrational s, are limits of periodic tiles of nearby parameters.
- In §9 we explain how to decompose Δ_s, in several ways, into pieces having bilateral symmetry. These pieces play an important role in our overall work. Establishing the bilateral symmetry requires to computer calculations, Calculations 1 and 2, which we do in Part V of the monograph.
- In §10 we prove the Main Theorem modulo 6 more calculations, Calculations $3, ..., 8$. We do these calculations in Part V.
- In §11 we establish various properties of the renormalization map R, such as its connection with continued fractions.
- In §12 we prove Theorem 1.2, Theorem 1.3, and Theorem 1.4. These results only require the Main Theorem and symmetry, so we present their proofs at the earliest possibly time. Theorem 1.2 is an immediate corollary of the stronger Theorem 12.1. Most of our effort in this chapter is devoted to proving Theorem 12.1.

CHAPTER 7

Elementary Properties

7.0.1. Notation. Basic Objects: Let (F_1, F_2, L_1, L_2) be the data that defines the octagonal PET at some parameter s, as shown in Figure 1.1. We repeat Figure 1.1 here for convenience. These objects all depend on s, but we usually suppress s from our notation. Here are the definitions of these sets:

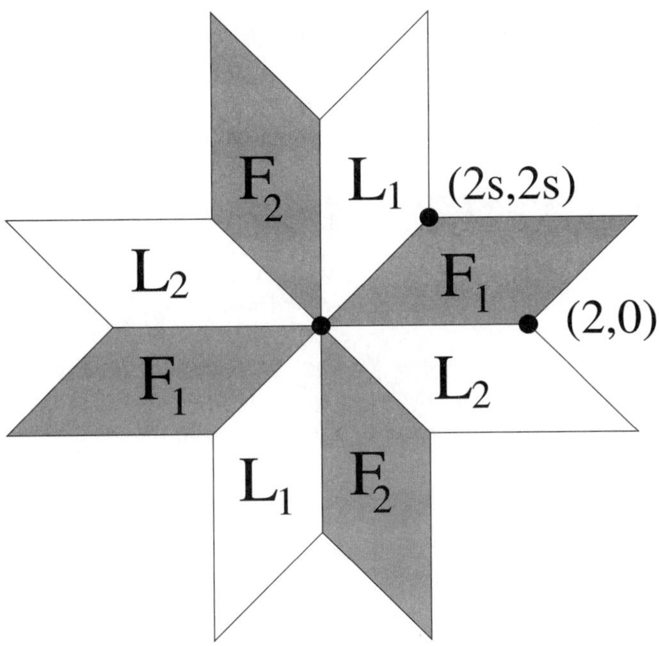

Figure 1.1: The scheme for the PET.

- F_1 is the parallelogram with vertices $(\epsilon_1 + \epsilon_2 s, \epsilon_2 s)$ for $\epsilon_1, \epsilon_2 \in \{-1, 1\}$.
- F_2 is the parallelogram with vertices $(\epsilon_2 s, -\epsilon_1 - \epsilon_2 s)$ for $\epsilon_1, \epsilon_2 \in \{-1, 1\}$.
- L_1 is the $\boldsymbol{Z}$ span of $(0, 2)$ and $(2s, 2s)$.
- L_2 is the $\boldsymbol{Z}$ span of $(2, 0)$ and $(2s, -2s)$.

Note that F_1 and F_2 differ from the labeled parallelograms in Figure 1.1 in that they have been translated so that they are centered at the origin.

We usually set $X = F_1$. The parallelogram X is the domain of the PET. We denote the PET by (X, f).

Dynamical Objects: Here are some objects objects associated to (X, f).

- Δ is the periodic tiling of the (X, f).
- Λ is the limit set of (X, f).
- Λ' is the aperiodic set of (X, f).

7.1. Intersection of the Parallelograms

As in the introduction, we define

$$O = F_1 \cap F_2. \tag{7.1}$$

This set has the following description:

- When $s \in (0, 1/2]$, O to be the square with vertices $(\pm s, \pm s)$.
- When $s \in (1/2, 1)$, O as the octagon with vertices

$$(\pm s, \pm(1-s)), \qquad (\pm(1-s), \pm s).$$

- When $s \in [1, \infty)$, O is the square with vertices $(\pm 1, 0)$ and $(0, \pm 1)$.

7.2. Intersection of the Lattices

LEMMA 7.1. *$L_1 \cap L_2 = \{0\}$ when s is irrational.*

Proof: A typical point in L_1 has the form $(2Bs, 2A + 2Bs)$ and a typical point in L_2 has the form $(2C + 2Ds, -2Ds)$ for $A, B, C, D \in \boldsymbol{Z}$. If two such points coincide, we have

$$Bs = C + Ds, \qquad A + Bs = -Ds. \tag{7.2}$$

Since s is irrational, the first equation is only possible if $C = 0$ and $B = D$. But the second equation gives $A = 2Ds$. This forces $A = D = 0$. In short both points are 0. □

Remark: We don't care about the intersection of the lattices in the rational case, but we mention that when $s = p/q$, both L_1 and L_2 contain the vectors $(2p, 0)$ and $(0, 2q)$. Hence $L_1 \cap L_2$ is a sub-lattice of both.

7.3. Rotational Symmetry

Now we discuss the symmetry of the system (X_s, f_s). Define

$$\iota(x, y) = (-x, -y). \tag{7.3}$$

Note that $\iota(X) = X$ for every parameter.

LEMMA 7.2 (Rotation). *ι and f_s commute for all $s \in (0, 1)$.*

Proof: ι preserves F_1, F_2, L_1, and L_2. For this reason ι commutes with f. □

7.4. Central Tiles

We make the following definitions.

- When $s \in (1/2, 1)$, we call $O = F_1 \cap F_2$ the *central tile*.
- When $s \leq 1/2$ or $s \geq 1$, the intersection $F_1 \cap F_2$ is a square. This square generates a grid in the plane, and finitely many squares in this grid lie in $X = F_1$. We call these squares the *central tiles*.

See Figure 7.1. Our definition is designed so that the complement of the central tile(s) plays the same role for all parameters of s.

Let X^0 denote the portion of X which lies to the left of the central tiles. We have

$$X = X^0 \cup \text{central tiles} \cup \iota(X^0). \tag{7.4}$$

Here ι is reflection in the origin, as defined in the previous section.

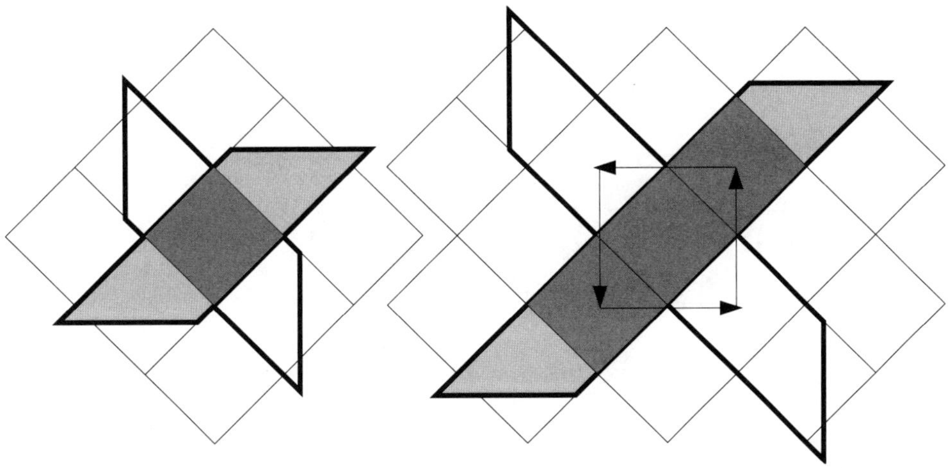

Figure 7.1: The central tiles (dark) for $s = 5/4$ and $t = 9/4$.

For any relevant subset $S \subset X$, we define

$$S^0 = S \cap X_0. \tag{7.5}$$

The subsets of Δ_s and Λ_s lying to the right of the central tiles are reflected images of Δ_s^0 and Λ_s^0. this reason, we will usually consider the picture just on the left hand side.

7.5. Inversion Symmetry

LEMMA 7.3 (Inversion). *(X_t, f_t) and (X_s, f_s^{-1}) are conjugate if $t = 1/2s$.*

Proof: Let L be the line through the origin which makes an angle of $\pi/8$ with the x-axis. Referring to Figure 1.1, the line L is an angle bisector for F_1. If we reflect Figure 1.1 in L and then dilate by $1/2s$, we get another instance of Figure 1.1, corresponding to the parameter t. Up to similarity, the corresponding PET for t is the inverse of the one for the parameter s. □

7.6. Insertion Symmetry

Figures 7.2 and 7.3 show our next result in action.

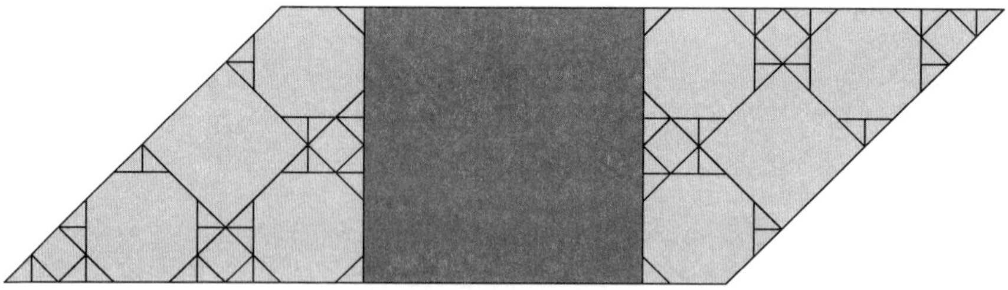

Figure 7.2: Δ_s for $s = 5/13$.

Figure 7.3: Δ_t for $t = 5/23$.

LEMMA 7.4 (Insertion). *Suppose* $s \geq 1$ *and* $t = s + 1$, *or suppose* $s \leq 1/2$ *and* $t = s/(2s+1)$. *The restriction of* f_s *to* $X_s^0 \cup \iota(X_s^0)$ *is conjugate to the restriction of* f_t *to* $X_t^0 \cup \iota(X_t^0)$. *The conjugacy is a piecewise similarity.*

Proof: The case when $s < 1/2$ is equivalent to the case $s > 1$ by the Inversion Lemma. So, we will take $s > 1$ and $t = s + 1$.

We consider how X_s and X_t sit relative to the grid of diamonds mentioned above. When $s, t > 1$, the squares in the grid are diamonds – their diagonals are parallel to $(\pm 2, 0) \in L_1$ and $(0, \pm 2) \in L_2$. There are two more diamonds contained in X_t than there are in X_s. On the central tile, the map f has the obvious action shown in Figure 7.1.

The sets X_s^0 and X_s^t have the same position relative to the diamond grid, and there is an obvious translation carrying the one set to the other. This translation extends to give piecewise translation from the complement of the diamonds in X_s to the complement of the diamonds in X_t. Call points related by this piecewise translation *partners*.

Let $p_s \in X_s^0$ and $p_t \in X_s^t$ be partners. Let λ_s and λ_t respectively be the vectors in $(L_2)_s$ and $(L_2)_t$ such that $p_s + \lambda_s \in (F_2)_s$ and $p_t + \lambda_t \in (F_2)_t$. We have either $\lambda_s = \lambda_t + (2, 0)$ or $\lambda_s = \lambda_t + (0, 2)$, depending on whether or not $p_s + \lambda_s$ and $p_t + \lambda_t$ lie in the top or bottom of $(F_2)_s$ and $(F_2)_t$ respectively. The answer (top/bottom) is the same for s as it is for t. In short, the two new points $p_s + \lambda_s$ and $p_t + \lambda_t$ are again partners. Repeating this argument, we see that $f_s(p_s)$ and $f_t(p_t)$ are partners. This is what we wanted to prove. □

7.7. The Tiling in Trivial Cases

LEMMA 7.5. Δ_s *consists entirely of squares and right-angled isosceles triangles when* $s = 1, 2, 3\ldots$ *and when* $s = 1/2, 1/4, 1/6, \ldots$.

Proof: We check this for $n = 1$ just by a direct calculation. See Figure 7.4.

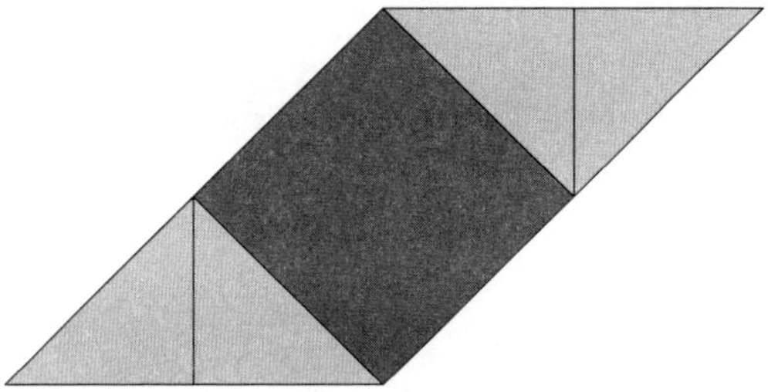

Figure 7.4: The tiling Δ_s for $s = 1$.

The cases $n = 2, 3, 4...$ now follow from the Insertion Lemma. The cases $n = 1/2, 1/4, 1/6...$ follow from the cases $n = 1, 2, 3...$ and the Inversion Lemma. $\square$

CHAPTER 8

Orbit Stability and Combinatorics

8.1. A Bound on Coefficients

We suppress the parameter s from our notation. Consider the set of points

$$S(V_1, V_2) = \{p \in F_1 |\ p + V_2 \in F_2, p + V_1 + V_2 \in F_1\}. \tag{8.1}$$

This set, if nonempty, is convex: It is the intersection of 3 parallelograms. Here $V_1 \in L_1$ and $V_2 \in L_2$. We write

$$V_1 = a_1(0,2) + b_1(2s, 2s), \qquad V_2 = a_2(2,0) + b_2(2s, -2s). \tag{8.2}$$

These coefficients are all integers. In this section we prove a technical lemma which we will use in the next section. Qualitatively, the lemma says that the vectors which arise in the sets $S(V_1, V_2)$, for $s \in [1/4, 1/\sqrt{2}]$, are short in an algebraic sense (as well as in a geometric sense).

LEMMA 8.1. *If $S(V_1, V_2)$ is nonempty and $s \in [1/4, 1/\sqrt{2}]$ then $|a_j| \leq 1$ and $|b_j| \leq 2$.*

Proof: We will give the proof for V_1. The proof for V_2 is similar. Inspecting the positions of the vertices of F_1 and F_2, we see that the maximum distance between a point of F_1 and a point of F_2 is

$$\lambda = \|(1, 2s+1)\| = \sqrt{4s^2 + 4s + 2}. \tag{8.3}$$

On the other hand, if $|a_1| \geq 2$, we have $\|V_1\| \geq 2\sqrt{2} > \lambda$. Hence $|a_1| \leq 1$. If $|b_1| \geq 4$ then $\|V_1\| \geq 8s > \lambda$. Hence $|b_1| \leq 3$. If $|b_1| = 3$ then $\|V_1\| \geq 6s > \lambda$ as long as $s > 1/3$.

So, we are left with the case $s \in [1/4, 1/3]$ and $|b_1| = 3$. In this case, we have the estimate

$$\|V_1\| \geq \min(\|(6s, 6s)\|, \|(6s, 6s-4)\|) =$$

$$\min(6\sqrt{2}s, \sqrt{16 - 48s + 72s^2}) > 8s. \tag{8.4}$$

The last estimate uses $s \leq 1/3$. From Equation 8.4, we see that $\|V_1\| > \lambda$ in this case as well. □

8.2. Sharpness

Recall from §2.4 that (X, f) defines a partition $\mathcal{A}_1$ consisting of the maximal domains on which f is defined and continuous. Each tile P of $\mathcal{A}_1$ has associated to it a vector V_P such that $f(p) = p + V_P$ for all $p \in P$. We call (X, f) *sharp* if $V_P = V_Q$ implies that $P = Q$. In other words, different tiles in the partition have different translations associated to them.

LEMMA 8.2. *If $s < 1$, then (X_s, f_s) is sharp unless $s = 1/2, 1/3, 1/4, ...$*

Proof: By Insertion Symmetry, it suffices to prove this result for $s \in [1/4, 1)$. By inversion symmetry, it suffices to prove the result for both f and f^{-1} when $s \in J = [1/4, 1/\sqrt{2}]$. We will prove the result for f when $s \in J$. The result for f^{-1} has essentially the same proof.

If (X_s, f_s) is not sharp, then we can find vectors $(V_1, V_2) \in L_1 \times L_2$ and $(V_1', V_2') \in L_1 \times L_2$ such that

- $S(V_1, V_2)$ and $S(V_1', V_2')$ are both nonempty,
- $V_1 + V_2 = V_1' + V_2'$
- $(V_1, V_2) \neq (V_1', V_2')$.

Let

$$W_1 = V_1 - V_1' = (2Bs, 2A + 2Bs) \in L_1,$$

$$W_2 = V_2' - V_2 \in L_2 = (2C + 2Ds, -2Ds) \in L_2. \tag{8.5}$$

We have $W_1 = W_2 \in L_1 \cap L_2$ and this gives the same equations as in Lemma 7.1. Solving Equation 7.2, we get

- $s = C/(B - D)$ if $B \neq D$
- $s = -2B/A$ if $B = D$.

Lemma 8.1 gives bounds

$$\|A\| \leq 2, \qquad \|B\| \leq 4, \qquad \|C\| \leq 2, \qquad \|D\| \leq 4. \tag{8.6}$$

The case $B = D$ yields no solutions in J.

Consider the case $B \neq D$. We have $|C| \leq 2$ and $|B - D| \leq 8$. The only solutions in J are $s = 1/2, 1/3, 1/4$ and $s = 2/3, 2/5, 2/7$. We check by hand that the system is sharp in the latter 3 cases. □

Remark: In fact, (X_s, f_s) is not sharp when $s = 1/n$. This will not bother us at all.

8.3. The Arithmetic Graph

Recall that $f' : (X_1 \cup X_2) \to (X_2 \cup X_1)$ is the square root of the PET map f. Suppose that $p_0 \in X$ is some point on which the orbit of f is well defined. Call this orbit $\{p_i\}$. There are unique vectors $V_i \in L_1$ and $W_i \in L_2$ such that

$$f'(p_i) = p_i + V_i, \qquad f'(p_i + V_i) = p_i + V_i + W_i = p_{i+1}.$$

We call the sequence $\{(V_i, W_i)\}$ the *symbolic encoding* of the orbit.

We define the *arithmetic graph* to be the polygon whose ith vertex is $V_0 + ... + V_i$. We define the *conjugate arithmetic graph* to be the polygon whose ith vertex is $W_0 + ... + W_i$. The arithmetic graph is a polygon with vertices in L_1 and the conjugate arithmetic graph is a polygon with vertices in L_2. We call these two objects together the *arithmetic graphs* of the orbit.

For each $j = 1, 2$ and each s there is a canonical isomorphism from L_j to $\boldsymbol{Z}^2$. The isomorphism sends the two basis vectors listed in §7.0.1 to the standard basis for $\boldsymbol{Z}^2$. These isomorphisms allow us to compare the graphs at different parameters. When we consider the images of the arithmetic graphs under this isomorphism, we will call them simply the *graphs*.

The arithmetic graph plays a major role in the analysis of outer billiards on kites in [**S1**]. Also, the arithmetic graphs associated to the octagonal PETs are quite pretty. The paper [**S5**] works them out in detail for the parameter $s = 1/\sqrt{2}$. In this monograph they play a more minor role.

LEMMA 8.3. *If s is irrational and p is a periodic point of f, then both the arithmetic graph of p and the conjugate arithmetic graph are closed polygons.*

When the orbit is periodic, of period n, we have

$$\sum_{i=1}^{n}(V_i + W_i) = 0, \qquad \sum_{i=1}^{n} V_i = -\sum_{i=1}^{n} W_i. \tag{8.7}$$

The first equation implies the second. But then the common sum belongs to $L_1 \cap L_2$. Lemma 7.1 gives us

$$\sum_{i=1}^{n} V_i = 0, \qquad \sum_{i=1}^{n} W_i = 0. \tag{8.8}$$

These two equations are equivalent to the lemma. □

8.4. Orbit Stability

We call a periodic point p *stable* relative to the parameter s if p is also periodic with respect to all parameters s' sufficiently close to s, and the integer arithmetic graphs associated to (p, s') are the same as those associated to (p, s). In this case, we say informally that small variations in the parameter do not destroy the orbit. Since each periodic orbit corresponds to a periodic tile, and *vice versa*, we can also speak about stable and unstable periodic tiles.

LEMMA 8.4. *A periodic point is stable if and only if its arithmetic graphs are closed loops.*

Proof: Let p be a periodic point, relative to the parameter s. Let $\{s_n\}$ be a sequence of parameters converging to s. Let N be the periodic of p relative to the parameter s. Suppose first that the arithmetic graphs of p are closed loops. No point of ∂X_s has a well-defined orbit. So, p lies in the interior of X_s. Hence p lies in the interior of X_{s_n} for all sufficiently large n. Moreover, by continuity, the first N iterates of f_{s_n} are defined on p once n is sufficiently large. For n sufficiently large, the first N steps of the arithmetic graphs of p relative to s_n must be the same as the first N steps of the integer arithmetic graphs of p with respect to s. But then these graphs form closed polygons for all s_n with n large enough. Hence p is periodic with respect to s_n and has the same integer arithmetic graphs.

Suppose on the other hand that the arithmetic graphs of p are not closed. Then we get relation in Equation 8.7, where each side is the same nontrivial vector in $(L_1)_s \cap (L_2)_s$. If p was stable then we would get a nontrivial element in the intersecton of the two lattices for all s' sufficiently close to s. But this is impossible when s' is irrational. Hence p is an unstable periodic point relative to s (and s is rational). □

As an immediate corollary, we have

LEMMA 8.5 (Stability). *All periodic orbits relative to an irrational parameter are stable.*

Remark: Theorem 1.4 characterizes unstable orbits as those whose periodic tiles are right-angled isosceles triangles.

8.5. Uniqueness and Convergence

To each periodic tile of Δ we associate the ordered pair of $\boldsymbol{Z}^2$ polygons which are the graphs associated to any point of the tile. We call these polygons the graphs associated to the tile.

LEMMA 8.6 (Uniqueness). *Suppose that $s < 1$ and $1/s \notin \boldsymbol{Z}$. Then distinct periodic tiles in Δ_s have distinct graphs.*

Proof: Suppose that p and q are both points having period n. We want to show that p and q lie in the same perodic tile of the partition $\mathcal{A}_n$ constructed in §2.4. Let $p_0 = p$ and $p_k = f_s^k(p_0)$. Likewise define q_k. There are vectors $V_1, ..., V_n$ so that $p_k = p_{k-1} + V_k$ and $q_k = q_{k-1} + V_k$ for all k. But then, since (X_s, f_s) is sharp, we see that p_k and q_k belong to the same tile of $\mathcal{A}_1$ for all k. But then p and q lies in the same periodic tile. □

LEMMA 8.7 (Approximation). *Let $s_\infty \in (0,1)$ be irrational, and let $\{s_n\}$ be a sequence of rationals converging to s_∞. Let P_∞ be a tile of Δ_∞. Then for all large n there is a tile P_n of Δ_n such that $\{\overline{P}_n\}$ converges to $\overline{P}_\infty$. The graphs associated to P_n are the same as the ones associated to P_∞.*

Proof: Let $\Gamma(z,t)$ denote the graph associated to a point or tile z at the parameter t. Let $p_1, ..., p_m$ be any finite collection of points in P_∞. By the Stability Lemma, $\Gamma(p_j, s_n) = \Gamma(P_\infty, s_\infty)$ for large n. By the Uniqueness Lemma, these points all lie in the same convex tile P_n. By choosing these points very close to the vertices of P_∞ we see that and subsequential limit of $\{\overline{P}_n\}$ contains $\overline{P}_\infty$, and moreover $\Gamma(P_n, s_n) = \Gamma(P_\infty, s_\infty)$ for large n.

Suppose that $\overline{P}_n$ does not converge to $\overline{P}_\infty$. Passing to a subsequence we can assume that $\{\overline{P}_n\}$ converges to some closed convex set $\overline{Q}_\infty$ which strictly contains $\overline{P}_\infty$. Almost every point of X_s has a well-defined f_s-orbit. So, we can find a point $q \in Q_\infty - \overline{P}_\infty$ which has a well-defined orbit relatice to s_∞. Note that $q \in P_n$ for all large n. If q is aperiodic, then the period of the f_{s_n} orbit of q tends to ∞ with n, contradicting $q \in P_n$. Suppose q is periodic. For large n we have

$$\Gamma(q, s_\infty) = \Gamma(q, s_n) = \Gamma(P_n, s_n) = \Gamma(P_\infty, s_\infty).$$

The first equality is the Stability Lemma. But then $q \in P_\infty$ by the Uniqueness Lemma. □

8.6. Ruling out Thin Regions

We say that a translation T of $\boldsymbol{R}^2$ is *adapted* to a parallelogram F if there are opposite sides e and e^* of F such that $T(e^*) \cap e$ is a segment.

LEMMA 8.8. *Suppose that $\{F_n\}$ is a convergent sequence of parallelograms and $\{L_n\}$ is a convergent sequence of lattices, and $\{W_n\}$ is a convergent sequence of vectors such that $W_n \in L_n$. Suppose that F_n is a fundamental domain for L_n for all n. If translation by W is adapted to F then translation by W_n is adapted to F_n for all large n.*

Proof: Let $\mathcal{F}_n$ denote the finite set of L_n translates of F_n which contain more than one point of ∂F_n. Let e_n and e_n^* be the edges of F_n which converge to e and e^* respectively. Since $\boldsymbol{R}^2$ is tiled by L_n translates of F_n, our hypotheses imply that $F_n + W_n$ is a member of $\mathcal{F}_n$ once n is sufficiently large. Our result follows immediately from this. □

Recall that f' is the square root of the map f. Let $J \subset (0,1)$ be some compact interval. We fix some parameter $s \in J$ and consider everything relative to this parameter. We call two parallel sequence *offset* if they do not lie in the same line. Let $\mathcal{A}$ be the partition for f'.

LEMMA 8.9. *For any $d > 0$ there is some $\epsilon > 0$ with the following property. If is impossible for two offset segments in the boundary of $\mathcal{A}$ to have length greater than d and to have their centers less than ϵ apart. The constant ϵ only depends on the interval J and on d.*

Proof: Suppose we had a sequence of counter-examples, and a corresponding sequence $\{s_n\} \subset J$. Passing to a subsequence, we can assume that $s_n \to s$. At the parameter s, the relevant line segments overlap. For ease of exposition, we suppress the dependence on the parameter s_n. Recall that the PET is given by (F_1, F_2, L_1, L_2). Let e and e^* be the offset seqments, and let q and q^* be their centers. Looking at the vectors associated to the regions on either side of e and e^*, we find distinct vectors $V, V^* \in L_2$ such that $q + V$ and $q^* + V^*$ lie in parallel edges of ∂F_2. But then we consider the parallelograms F_2 and $F_2 + W \in L_2$, where $W = V - V^*$. These two parallelograms have two sides which are parallel, very close together, and offset. Letting $n \to \infty$, we contradict Lemma 8.8. □

LEMMA 8.10. *For any $d > 0$ there is some $\epsilon > 0$ with the following property. Suppose that P is a periodic tile for f and some edge e of P has length greater than d and does not lie in an edge of $\mathcal{A}$. Then e must be at least ϵ from any parallel edge in the boundary of $\mathcal{A}$. The constant ϵ only depends on the interval J and on d.*

Proof: There is some smallest k so that $(f')^k$ is not defined on e. If $k \geq 2$, we can replace P by the tile containing $Rf'(P)$ and e by the edge of P' containing $Rf'(e)$. Here $R : F_2 \to F_1$ is rotation. The new pair (P', e') has exactly the same relevant features as (P, e), except that perhaps the length d has increased.

Thus, without loss of generality, we can consider the case when $k = 1$. That is, f' is not defined on e. But then e is contained in an edge of $\mathcal{A}$ which is parallel to, offset from, and close to, another edge of $\mathcal{A}$. This contradicts Lemma 8.9 once the constants are appropriately chosen. □

8.7. Joint Convergence

Suppose that s is an irrational parameter and that P and P^* are tiles of Δ_s which share an edge. If $\{s_n\}$ is a sequence which converges to s, we know by Lemma 8.7 that there are tiles P_n and P_n^* of Δ_n such that $P_n \to P$ and $P_n^* \to P^*$. We don't know yet that P_n and P_n^* also share an edge. Our main goal is to prove the following result:

LEMMA 8.11. *$P_n \cap P_n^*$ is a segment e_n, and e_n converges to $e = P \cap P^*$.*

We will prove this result through a series of smaller lemmas.

LEMMA 8.12. *P_n has an edge f_n which converges to e.*

Proof: P_n has at most 8 sides, and the possible slopes for these sides lie in $\{-1, 0, 1, \infty\}$. The lemma follows almost immediately from this fact, and from the fact that $\{\overline{P}_n\}$ converges to $\overline{P}$. □

The same result holds for P_n^*. Moreover, given the constraints on the slopes, the edges f_n and f_n^* are parallel for large n. Let $e_n = f_n \cap f_n^*$. For all we know now, e_n is the empty set. We want to show that e_n is nonempty and converges to e.

Now we use the notation from the previous section.

LEMMA 8.13. *It suffices to prove Lemma 8.11 in the case when e is contained in edge of the partition for f'_s.*

Proof: This argument is just like the one used in Lemma 8.10. Note that there is some smallest k so that $(f'_s)^k(e)$ lies in the partition for f'_s. If $k > 1$, then both tiles P and P^* lie in the interior of some open tile in the partition for f'_s. But then, by continuity, both P_n and P_n^* lie in the corresponding tile in the partition for f'_{s_n}. Now we replace P and P^* by $Rf'_s(P)$ and $Rf'_s(P^*)$ and we replace P_n and P_n^* by $Rf'_{s_n}(P_n)$ and $Rf'_{s_n}(P_n^*)$. Here $R : F_2 \to F_1 = X$ is 90 degree rotation. Once we make these replacements, we get a new counterexample where k has been decreased by 1. We continue this construction until we arrive at a counterexample with $k = 0$. □

Since P and P^* are contained in distinct tiles of the partition $\mathcal{A}$ for f'_s, the tiles P_n and P_n^* are contained in distinct tiles of the partition $\mathcal{A}(n)$ for f'_{s_n}. Moreover, there is an edge of $\mathcal{A}(n)$ which is close to e. Hence P_n has a long edge f_n which is parallel to, and close to, an edge of $\mathcal{A}(n)$. Once n is sufficiently large, Lemma 8.10 says that f_n must lie in this edge of $\mathcal{A}(n)$. But the same argument works for f_n^*. Hence f_n and f_n^* lie in the same line. Since both f_n and f_n^* converge to e, this is only possible if $f_n \cap f_n^*$ also converges to e. This completes the proof of Lemma 8.11.

CHAPTER 9

Bilateral Symmetry

9.1. Pictures

The kind of symmetry described in this chapter looks obvious from the pictures, but it seems to require a nontrivial computational proof. In view of the Inversion and Insertion symmetry, we will describe the symmetry just for $s \in [1/4, 1)$. In this section we will show pictures and in the next section we will describe things in terms of formulas. Ultimately, we depict two different decompositions

$$\Delta_s^0 = A_s \cup B_s = P_s \cup Q_s \tag{9.1}$$

into regions which have bilateral symmetry. Here are typical pictures for the case $s \in (1/4, 1/2)$.

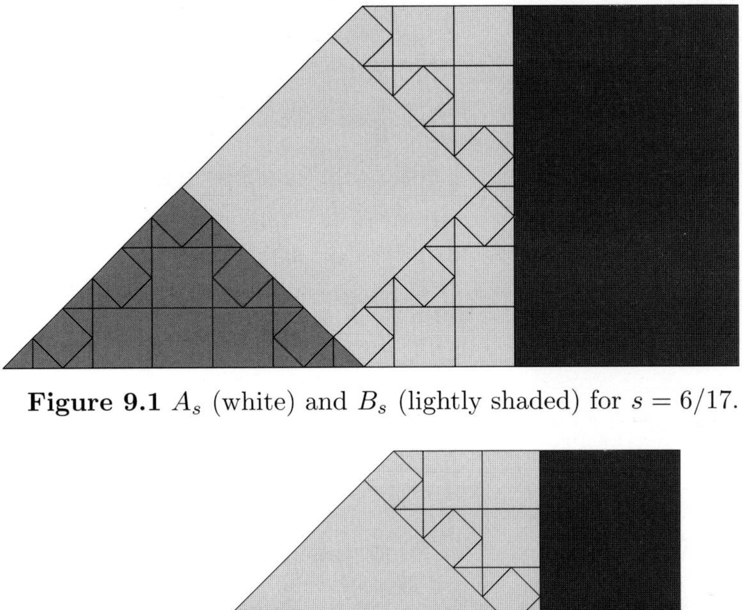

Figure 9.1 A_s (white) and B_s (lightly shaded) for $s = 6/17$.

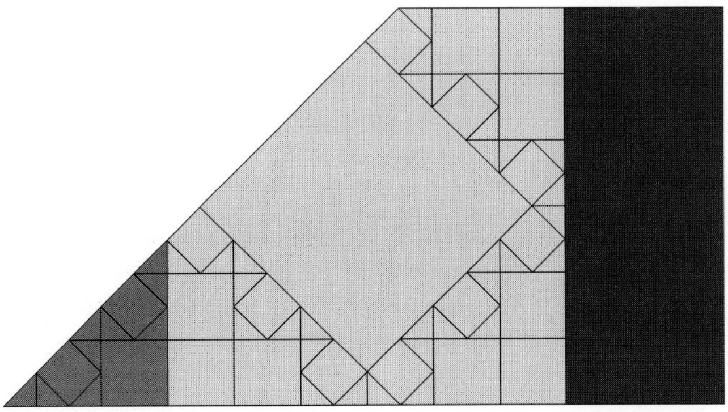

Figure 9.2 P_s (white) and Q_s (lightly shaded) for $s = 6/17$.

Here are typical pictures for the case $s \in (1/2, 1)$.

Figure 9.3 A_s (white) and B_s (lightly shaded) for $s = 8/13$.

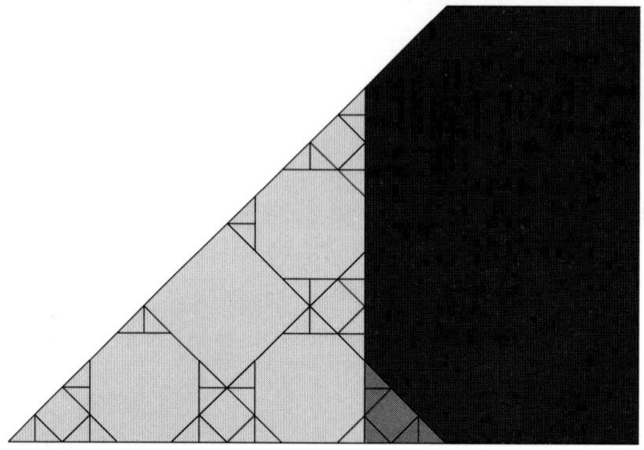

Figure 9.4 P_s (white) and Q_s (lightly shaded) for $s = 8/13$.

9.2. Definitions and Formulas

We say that a line L is a *line of symmetry* for Δ_s if

$$\Delta_s \cap \Big(X_s \cap R_L(X_s)\Big) \tag{9.2}$$

is invariant under the reflection R_L in L. Note that X_s itself need not be invariant under R_L. Consider the following lines:

- Let H be the line $y = 0$.
- Let V be the line $x = -1$.
- For D_s be the line of slope -1 through $(-s, -s)$.
- For $s \in [1/4, 1/2]$, E_s be the line of slope -1 through $(-3s, -3s)$.
- For $s \in [1/2, 1]$, E_s is the line of slope 1 through $(-s, -s)$.

We call these the *fundamental lines of symmetry*. For each line L above, the line $\iota(L)$ is also a line of symmetry.

We define

$$A_s = (X_s \cap R_H(X_s))^0, \qquad B_s = X_s \cap R_V(X_s),$$

$$P_s = X_s \cap R_D(X_s), \;\; Q_s = X_s \cap R_E(X_s). \tag{9.3}$$

Here is the main result in this chapter:

LEMMA 9.1 (Bilateral). *For all* $s \in [1/4, 1]$, *the lines* H, V, D, E *respectively are the lines of symmetry for the regions* A, B, P, Q.

We prove this result modulo two computer calculations, which we defer until Part V.

We first deal with the pieces A and B. We have the decomposition

$$X' := X - \text{central tiles} = A \cup B \cup \iota(A) \cup \iota(B). \tag{9.4}$$

Each of the pieces on the right hand side has a vertical line of bilateral symmetry. The reflections across these vertical lines gives rise to a piecewise isometry $\mu : X' \to X'$. Note also that X' is an f_s-invariant set. If $p \in A$ we define $\mu(p)$ to be reflection in the vertical line of symmetry for A, etc. In Part V we prove by computer calculation the following result.

LEMMA 9.2 (Calculation 1). *If* $s \in [1/4, 1]$ *then* $\mu_s \circ f_s \circ \mu_s = f_s^{-1}$ *wherever both maps are defined. Moreover, both* A_s *and* B_s *are clean sets.*

By rotational symmetry, it suffices to prove that $\Delta \cap A$ and $\Delta \cap B$ are μ-invariant. We do this for A. The proof for B is the same.

Let τ be a tile of Δ that is contained in A. We just need to see that $\mu(\tau)$ is a periodic tile for Δ. Indeed, it suffices to show that $\mu(\tau)$ is contained in a periodic tile of Δ, because μ is an involution. Let n be the period of f on τ. We just need to show that the first n iterates of f are defined on all points of $\mu(\tau)$

Let $\Gamma(p, f)$ denote the arithmetic graph associated to the map f relative to the point p. Note that $\mu f \mu$ is defined on almost all points of τ, and equals f^{-1} on such points. Hence, by Calculation 1, the map f is n-periodic on almost all points of $\mu(\tau)$. If one of the first n iterates of f is not defined on $\mu(\tau)$ then several distinct periodic tiles intersect $\mu(\tau)$. By the Uniqueness Lemma from the previous chapter, there are points $q_1, q_2 \in \mu(\tau)$ such that

$$\Gamma(q_1, f) \neq \Gamma(q_2, f).$$

Set $p_j = \mu(q_j) \in \tau$. Tautologically, we have

$$\Gamma(p_1, \mu f \mu) \neq \Gamma(p_2, \mu f \mu).$$

But both maps $\mu f \mu$ and f^{-1} are defined on p_j. Hence

$$\Gamma(p_j, \mu f \mu) = \Gamma(p_j, f^{-1}),$$

by Calculation 1. Hence

$$\Gamma(p_1, f^{-1}) \neq \Gamma(p_2, f^{-1}).$$

But $\Gamma(p_j, f)$ determines $\Gamma(p_j, f^{-1})$, and vice versa. Hence

$$\Gamma(p_1, f) \neq \Gamma(p_2, f).$$

This contradicts the fact that $p_1, p_2 \in \tau$.

Now we turn to the second partition. We have a decomposition just like Equation 9.4, but with P and Q replacing A and B. This time, each piece in the partition has a line of slope -1 as a line of symmetry. We define ν with respect to these lines just as we defined μ above.

In Part V we prove the following result by direct calculation.

LEMMA 9.3 (Calculation 2). *If* $s \in [1/4, 1]$, *then* $\nu_s \circ f_s \circ \nu_s = f_s^{-1}$ *wherever both maps are defined. Moreover, both* P_s *and* Q_s *are clean sets.*

The rest of the proof is identical to what we did for the first partition.

Remark: Calculations 1 and 2 involve finitely many integer linear algebra operations performed on finitely many convex integer polyhedra in $\boldsymbol{R}^3$. Were we to take $s \in (0,1)$ rather than $s \in [1/4,1)$ we would have to contend with infinitely many polyhedra, and we could not make a direct finite calculation. The Insertion Symmetry is what allows us to reduce the desired calculations to finite ones.

CHAPTER 10

Proof of the Main Theorem

10.1. Discussion and Overview

The Inversion Lemma and the Insertion Lemma go part of the way towards proving the Main Theorem. These two results give the desired relation between (X_s, f_s) and (X_t, f_t) for pairs $(s, 1/2s)$, and for pairs $(s, s+1)$ when $s > 1$. However, these results are not strong enough to establish the Main Theorem. For instance, when $s = 2/5$ we have $R(s) = 1/4$. The relations above give no connection between $5/4 = 1/(2s)$ and $1/4$.

In this chapter, we reduce the Main Theorem to some computer calculations, which we call Calculations 3,...,8. We do these calculations in Part V. For any given parameter, an obvious finite proof of the Main Theorem suggests itself. Find the partitions associated to each of the maps and verify the result. If things are set up carefully, we need only check a single point in each tile of the partitions. Calculations 3-8 treat the whole 1-parameter family as a 3-dimensional system and then make a direct verification along the lines suggested.

Here are the two results we prove in this chapter. The first is equivalent to the half of the Main Theorem corresponding to $s \in (0, 1/2)$. We use the notation from the Main Theorem. $Y_{\overline{s}}$ is the complement of the central tile(s) in the domain $X_{\overline{s}}$.

LEMMA 10.1. *Suppose $s \in (1, 2)$ and $\overline{s} = s - 1$. Let $\phi_s : Y_{\overline{s}} \to X_s$ be the map which is a translation on each half of $Y_{\overline{s}}$ and maps the acute vertices of $Y_{\overline{s}}$ to the acute vertices of X_s. Let $Z_s = \phi_s(Y_{\overline{s}})$. Then ϕ_s conjugates $f_{\overline{s}}|Y_{\overline{s}}$ to $f_s|Z_s$, and Z_s is a clean set. Either half of ϕ_s extends to the trivial tile of $\Delta_{\overline{s}}$ and maps it to tiles τ_1 and τ_2. The only nontrivial f_s-orbits which miss Z_s are contained in $\tau_1 \cup \tau_2$ and have period 2.*

Our other result is just a restatement of the half of the Main Theorem corresponding to $s \in (1/2, 1)$.

LEMMA 10.2. *Suppose $s \in (1/2, 1)$ and $\overline{s} = 1 - s$. Let $\phi_s : Y_{\overline{s}} \to X_s$ be the map which is a translation on each half of $Y_{\overline{s}}$ and maps the acute vertices of $Y_{\overline{s}}$ to the acute vertices of X_s. Let $Z_s = \phi_s(Y_{\overline{s}})$. Then ϕ_s conjugates $f_{\overline{s}}|Y_{\overline{s}}$ to $f_s^{-1}|Z_s$, and Z_s is a clean set. All nontrivial f_s-orbits intersect Z_s.*

Proof of the Main Theorem: Lemma 10.2 is just a restatement of the Main Theorem for $s \in (1/2, 1)$. Suppose that $s < 1/2$. By the Insertion Lemma, it suffices to consider the case when $s \in (1/4, 1/2)$. By the Inversion Lemma, the system (X_s, f_s) is conjugate to the system (X_t, f_t^{-1}), where $t = 1/2s$. Here $t \in (1, 2)$. But now Lemma 10.1 applies to the pair $(t, t-1)$ and $t - 1 = R(s)$. When we combine the conjugacy given by the Inversion Lemma with the one given by Lemma 10.1, we get the Main Theorem. □

A direct computational proof of Lemmas 10.1 is difficult. Consider, for instance, what happens as $s \to 1$. In this case, $\overline{s} \to 0$ and the area of $Y_{\overline{s}}$ tends to 0. But then the area of Z_s tends to 0 as well. On the other hand, the area of X_s does not tend to 0. Hence, the proportion of X_s taken up by Z_s tends to 0. But then the amount of time it takes for some orbits to return to Z_s probably tends (and, in fact, does tend) to ∞. A similar problem happens in Lemma 10.2 when $s \to 1$.

Our first two calculations stay away from the parameters which lead to these unbounded calculations.

LEMMA 10.3 (Calculation 3). *Lemma* 10.1 *holds for all* $s \in [5/4, 2]$.

LEMMA 10.4 (Calculation 4). *Lemma* 10.2 *holds for all* $s \in [1/2, 3/4]$.

Let

$$T(s) = \frac{s-2}{2s-3}. \tag{10.1}$$

Below we prove the following results.

LEMMA 10.5 (Induction I). *If Lemma* 10.1 *is true for some* $u \in (1, 3/2)$, *then Lemma* 10.1 *is also true for* $s = T^{-1}(u)$.

LEMMA 10.6 (Induction II). *If Lemma* 10.2 *is true for some* $u \in (1/2, 1)$, *then Lemma* 10.1 *is also true for* $s = T(u)$.

The map T has the action $1/2 \to 3/4 \to 5/6 \to \cdots$ The map T^{-1} has the action $3/2 \to 5/4 \to 7/6 \to \cdots$. Hence

$$\bigcup_{k=1}^{\infty}(1, 3.2) = T^{-k}[5/4, 3/2), \qquad (3/4, 1) = \bigcup_{k=1}^{\infty} T^k(1/2, 3/4). \tag{10.2}$$

Lemma 10.1 follows from Calculation 3, the Induction Lemma I, Equation 10.2. Likewise Lemma 10.2 follows from Calculation 4, the Induction Lemma II, and Equation 10.2.

10.2. Proof of Induction Lemma I

Let $u \in (1, 3/2)$ and $s = T^{-1}(u)$. Define $\omega_s : Y_u \to X_s$ by the formula

$$\omega_s(x, y) = (3 - 2s)(x, y) \pm (2 - 2s, 0). \tag{10.3}$$

The $(+)$ option for ω_s is taken when $x < 0$ and the $(-)$ option is taken then $x > 0$. Let

$$W_s = \omega_s(Y_u). \tag{10.4}$$

The left-hand side of Figure 10.1 shows W_s^0 (red) for $s = 22/19$. The right hand side of Figure 10.1 shows Y_u^0 (red) for $u = T(22/19) = 16/13$. Note that we have used two different shades of red (and blue) in these pictures, for the purpose of distinguishing certain subsets.

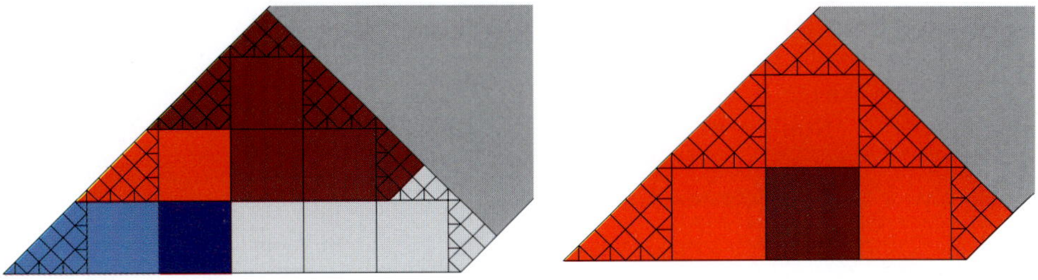

Figure 10.1: Some sets for $s = 22/19$ (left) and $u = T(s) = 16/13$ (right).

LEMMA 10.7 (Calculation 5). *Let $s \in (1, 4/3]$ and $u = T(s)$. Then ω_s conjugates $f_u|Y_u$ to $f_s|W_s$. Moreover, Z_s is a clean set.*

We define some auxilliary sets:

- The blue set in Figure 10.1 is isometric to Y_{s-1}, and we call it Z_s^0.
- Let δ_s be the diagonal vector pointing from the bottom left edge of Z_s^0 to the top left edge of Z_s^0.
- Let $Z_s = Z_s^0 \cup \iota(Z_s)$. Here ι is reflection in the origin.
- Let τ_u denote the square whose left side coincides with the right side of Z_u^0. The square τ_u is a darker red than the others in Figure 10.1, right.
- We define τ_s just as we defined τ_u, with s in place of u.

LEMMA 10.8 (Calculation 6). *Let $s \in (1, 5/4]$. Then*

(1) *τ_u is a tile of Δ_u, having period 2.*
(2) *$f_s^{-1}(p) = p + \delta_s$ for all $p \in Z_s^0$.*
(3) *$f_s^{-1}(X_s - Z_s - W_s) \subset Z_s \cup \tau_s \cup \iota(\tau_s)$.*

Now we prove the Induction Lemma I. This is really a diagram chase. Let $\overline{s} = s - 1$ and $\overline{u} = u - 1$. We are going to fill in the arrows of the following diagram.

$$\begin{array}{ccc} Y_{\overline{u}} & \xrightarrow{\phi_u} & Z_u \\ \updownarrow & & \updownarrow \\ Y'_{\overline{s}} & \longrightarrow & Z'_s \\ \downarrow & & \downarrow \\ Y_{\overline{s}} & \longrightarrow & Z_s \end{array} \tag{10.5}$$

The top row of the diagram is what we get from Lemma 10.1 for the parameter u. The bottom row will give us Lemma 10.1 for the parameter s.

Top Left Arrow: To explain the top left arrow, we compute $\overline{s} = \overline{u}/(2\overline{u} + 1)$. That is, $\overline{s}$ and $\overline{u}$ are related exactly as in the Insertion Lemma. The set Y'_u is obtained from Y_u by chopping off the rightmost central square from the left half of Y_u and the leftmost central square from the right half. The map ($\updownarrow$) is the conjugacy guaranteed by the Insertion Lemma.

Bottom Left/Right Arrows: The set Z'_s is obtained from Z_s just as Y'_s is obtained from Y_s. In Figure 10.1, the set Z'_s is the light blue set. The bottom left and bottom right arrows are both inclusion.

Top Right Arrow: The top right arrow is

$$f_s \circ \omega_s : Z_u \to Z'_s. \tag{10.6}$$

Here ω_s is the map from Calculation 5, and f_s is the PET on X_s, discussed in Calculation 6. The point here is that $\omega_s(Z_u)$ abuts the left edge of X_s^0, but is a translate of Z_s' by the vector δ_s: It is the light red set in Figure 10.1. Then, f_s translates $\omega_s(Z_u)$ to Z_s'. Statement 2 of Calculation 6 implies f_s conjugates $f_s|\omega_s(W_u)$ to $f_s|Z_s'$. Calculation 5 says that ω_s conjugates $f_u|Z_u$ to $f_s|\omega_s(Z_u)$. Hence, the map in Equation 10.6 conjugates $f_u|W_u$ to $f_s|Z_s'$.

Middle Arrow: The middle arrow is then defined to make the top square commute. At this point, we have the desired conjugacy from Lemma 10.1, for the parameter s, except that the (primed) sets are slightly too small.

Bottom Arrow: It follows from Statement 2 of Calculation 6 that the union of two square tiles in $Z_s - Z_s'$ are in the same orbit and have period 2. This lets us define the bottom right arrow to make the bottom square commute. We call this map ϕ_s and note that it satisfies the main conclusion (the conjugacy) in Lemma 10.1.

Extension to the Central Tile: Statement 1 of Calculation 6 guarantees that each half of ϕ_s extends to the central tile of $\Delta_{\overline{s}}$. The image of this central tile under the left half is τ_s and the image under the right half if $\iota(\tau_s)$.

The following lemma takes care of the one piece of unfinished business.

LEMMA 10.9. *Any nontrivial f_s-orbit, except those contained in $\tau_s \cup \iota(\tau_s)$, intersects Z_s.*

Proof: Consider first the orbit of a point $p \in W_s$. Let $q = \omega_s^{-1}(p)$. Since the Main Theorem is true for the parameter u, the orbit of q intersects Z_u^*. But then, by Calculation 5, the orbit of q intersects $\omega_s(Z_u^*) = f_s^{-1}(Z_s)$. But then $f_s(q) \in Z_s$. Finally, consider the orbit of $p \in X_s - Z_s - W_s$. If $p \in \tau_s \cup \iota(\tau_s)$, there is nothing to prove. Otherwise, Statement 3 of Calculation 6 finishes the proof. □

10.3. Proof of Induction Lemma II

We set $U = T^{-1}$. The proof of the Induction Lemma II is similar to the proof of the Induction Lemma I. Mainly we explain the differences. We define

$$\omega_s(x, y) = (2s-1)(x, y) \pm (2s-2). \tag{10.7}$$

We set $W_s = \omega_s(Y_u)$.

The left hand side of Figure 10.2 shows W_s^0 (red) and Z_s^0 (blue) for the parameter $s = 28/31$. The right hand side of Figure 10.2 shows Y_u^0 (red) and Z_u (light red) for $u = U(28/31) = 22/25$.

Figure 10.2: Some sets for $s = 28/31$ and $u = U(s) = 22/25$.

The rest of the definitions are done as in the previous section. The main difference here is that the tiles τ_s and τ_u belong to Z_s and Z_u respectively. This causes slightly differences in the statements of the two calculations below, which parallel Calculations 5 and 6.

LEMMA 10.10 (Calculation 7). *Let $s \in [3/4, 1)$ and $u = U(s)$. Then ω_s conjugates $f_u|Y_u$ to $f_s|W_s$. Moreover, Z_s is a clean set.*

LEMMA 10.11 (Calculation 8). *Let $s \in [3/4, 1)$. Then*

(1) *τ_s is a tile of Δ_s, having period* 2.
(2) *$f_s(p) = p + \delta_s$ for all $p \in (Z'_s)^0$.*
(3) *$f_s(X_s - Z_s - W_s) \subset Z'_s$.*

Here Z'_s has the same meaning as above.

The rest of the proof of Lemma 10.2 is like what we did for Lemma 10.1, but there are two small differences. First, this time we know that τ_s is a tile of Z_s, so all the nontrivial orbits intersect Z_s. Second, when $s \in [3/4, 5/6)$ and $u \in [1/2, 3/4)$ the tile τ_u does not exist. (This is why Statement 1 of Calculation 8 refers to τ_s rather than τ_u.) This does not change the argument in any substantive way. Figure 10.3 shows the same sets as in Figure 10.2 for $s = 4/5$ and $u = U(s) = 2/3$.

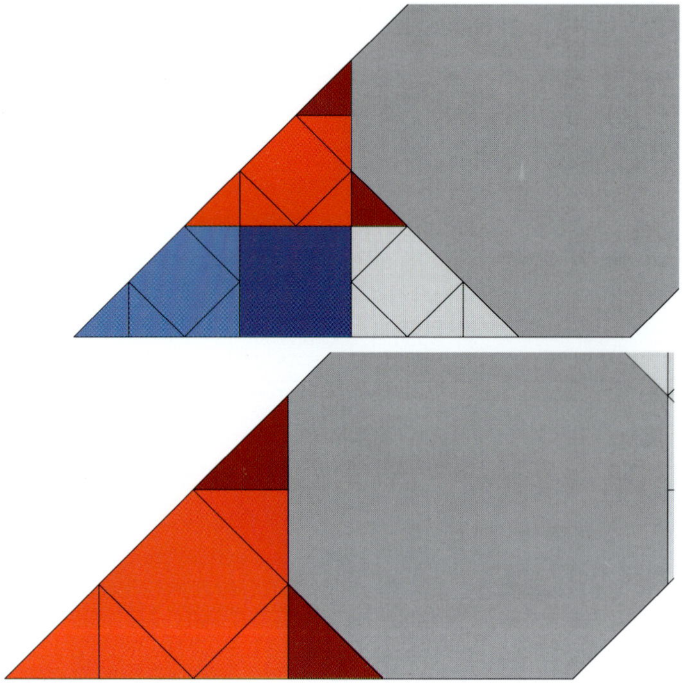

Figure 10.3: Some sets for $s = 4/5$ and $u = U(s) = 2/3$.

CHAPTER 11

The Renormalization Map

11.1. Elementary Properties

In this chapter we work out properties of the renormalization map R. Here we recall the definition $R : (0,1) \to [0,1)$ as follows:

- $R(x) = 1 - x$ if $x > 1/2$
- $R(x) = 1/(2x) - \text{floor}(1/(2x))$ if $x < 1/2$.

LEMMA 11.1. *Let $p/q \in (0,1) \cap \boldsymbol{Q}$. There is some k such that $R^k(p/q) = 0$.*

Proof: We write $p_k/q_k = R^k(p/q)$. If $p/q < 1/2$ then $2p < q$ and hence $q_1 < q_0$. If $p/q > 1/2$ then $q_1 = q_0$, and the previous argument shows that $q_2 < q_1$. So, in both cases, we have $q_2 < q_0$. Repeated applications of R^2 decrease the denominator, until we reach 0. □

For $s \in (0,1)$ we define $E(s)$ by the equation

$$\frac{1}{E(s)+1} \leq s \leq \frac{1}{E(s)}, \tag{11.1}$$

and (if necessary) the condition that $E(s)$ is even. What we mean here is that $E(s)$ is completely unambigious unless $s = 1/n$. In this case, the inequalities pin one of two choices of $E(s)$, and we choose the even one. That is,

$$E\Big(\frac{1}{2n}\Big) = E\Big(\frac{1}{2n+1}\Big) = 2n, \qquad n = 1,2,3... \tag{11.2}$$

We call s *trivial* is $s = 1/n$ for some integer $n = 2,3,4....$

LEMMA 11.2. *Suppose s is nontrivial and let $t = R(s)$.*

- *If $E(s) = 1$ then $E(t) \geq 2$.*
- *If $E(s) \geq 2$ then $E(t) = 1$ if and only if $E(s)$ is odd.*

Proof: This is an easy exercise. □

11.2. The Even Expanson

We define the *even expansion* of $s \in (0,1)$ to be the sequence

$$\{E(R^n(s))\}_{n \geq 0}. \tag{11.3}$$

We sometimes write $e_k = E(R^k(s))$ when the choice of s is either clear or unimportant. When s is rational, we stop the sequence at the point when $R^n(s) = 1/2k$ for some integer k. Thus, the even expansion of a rational number always ends in an even integer. By Lemma 11.2, we have the following structure:

- $e_k = 1$ implies that $e_{k+1} \geq 2$.
- $e_k = 2, 4, 6, ...$ implies that $e_{k+1} \geq 2$.
- $e_k = 3, 5, 7, ...$ implies that $e_{k+1} = 1$.
- The even expansion of a rational cannot end in $1, 2$.

Other than these restrictions, any sequence is possible. To see this, one simply starts with a point in the desired interval, and starts pulling back using the desired branches of R^{-1}. Here are some examples:

- $1/3$ has even expansion $(2, 2)$.
- $11/28$ has even expansion $(2, 3, 1, 6)$. The successive images under R are $3/11$, $5/6$, $1/6$.
- $-1/2 + \sqrt{3}/2$ has even expansion $(2, 2, 2, ...)$.
- $\sqrt{2}/2$ has even expansion $(1, 3, 1, 3, ...)$.
- $-1/2 + \sqrt{5}/2$ has even expansion $(1, 2, 3, 1, 2, 3, ...)$.

Below we will explain how to relate the even expansion of a number to (some version of) its continued fraction expansion.

11.3. Oddly Even Numbers

When s is rational, we tweak the definition of the continued fraction of a rational number, as follows. If the continued fraction fraction ends $(..., k)$, with $k > 1$ odd, we change it so that it ends $(..., k-1, 1)$. This does not change the value of the expression, when it is written out as a continued fraction. We call $s \in (0, 1)$ *oddly even* if the following properties hold.

(1) s has C.F.E. $(0, a_1, a_2, ...)$ with a_k even for all odd k.
(2) The even expansion of s is entirely even.
(3) $R^k(s) < 1/2$ for all $k = 0, 1, 2, ...$

Lemma 11.2 shows that Properties 2 and 3 are equivalent. Now we show that Properties 1 and 3 are equivalent.

LEMMA 11.3. *Let $(a, b, ...)$ be the first two terms in the even expansion of s. If a and b are both even, then $R^2(s) = G^2(s)$.*

Proof: Let $t = R(s)$ and $u = R(t)$. We compute

$$t = \frac{1}{2s} - \frac{a}{2}, \qquad u = \frac{1}{2t} - \frac{b}{2}. \tag{11.4}$$

But then, after some algebra, we get

$$s = \cfrac{1}{a + \cfrac{1}{(b/2) + u}}. \tag{11.5}$$

On the other hand, from Lemma 11.4 and the equation for the Gauss map, we see that $G(s) = 2t$ and $G(2t) = u$. Hence $R^2(s) = G^2(s)$. □

Lemma 11.3 immediately shows that Property 1 implies Property 3. Conversely, suppose that Property 1 fails. Applying Lemma 11.3 finitely many times if necessary, we reduce to the case where the first term in the continued fraction expansion of s is odd. If $s > 1/2$ we are done. Otherwise, s lies in one of the intervals $(1/3, 1/2)$, $(1/5, 1/4)$, $(1/7, 1/6)$,..., and R maps each of these intervals onto $(1/2, 1)$, by Lemma 11.2. Hence Property 3 fails.

11.4. The Even Expansion and Continued Fractions

We begin with a result in the oddly even case.

LEMMA 11.4. *s has even expansion $(2a_1, 2a_2, 2a_3, \ldots)$ if and only if s has continued fraction expansion $(0, 2a_1, a_2, 2a_3, a_4, \ldots)$.*

Proof: In view of Lemma 11.3, we just have to check this for the first two terms. Equation 11.5 implies the result for the first two terms. □

In general, the same method of proof establishes a translation between the even expansion and the signed continued fraction when $s < 1/2$. We illustrate the translation for (the randomly chosen) $s = 3073/7256$.

- We convert the even expansion to a new sequence by putting a 0 in front and then replacing each fragment $\ldots 2m-1, 1, \ldots$ with the decorated number $\overline{2m}$. For instance $(2,5,1,4,6,7,1,3,1,6) \to (0,2,\overline{6},4,6,\overline{8},\overline{4},6)$.
- We take our decorated sequence and cut every other term in half, starting with the 0. For instance, $(0,2,\overline{6},4,6,\overline{8},\overline{4},6) \to (0,2,\overline{3},4,3,\overline{8},\overline{2},6)$.
- We take the last sequence and convert it into a continued fraction and then put $(-)$ signs in front of all the decorated terms.

$$\frac{3073}{7256} = \cfrac{1}{2 + \cfrac{1}{3 - \cfrac{1}{4 + \cfrac{1}{3 + \cfrac{1}{8 - \cfrac{1}{2 - \cfrac{1}{6}}}}}}}$$

- To get a signed continued fraction in the sense discussed in the previous section, we put a $(-)$ sign in front of each term of our last sequence if and only if there are an odd number of terms preceding it, and then we remove the decorations. Thus $(0,2,\overline{3},4,3,\overline{8},\overline{2},6) \to (0,2,3,-4,-3,-8,2,-6)$.

11.5. Diophantine Approximation

Let E.E. stand for *even expansion*. Suppose $\sigma, s \in (0, 1/2)$ are such that σ is rational and s is irrational. We write $\sigma \to s$ if one of two situation holds.

(1) The E.E. of σ is an initial portion of the E.E. of s.
(2) The E.E. of σ has the form $(a_1, \ldots, a_{n-1}, 2a_n)$ and the E.E. of s has the form $(a_1, \ldots, a_{n-1}, 2a_n - 1, 1, \ldots)$.

For example, if s has E.E. $(3,1,3,1,3,1,\ldots)$ then $\sigma \to s$ when σ has E.E. $(3,1,4)$ or $(3,1,3,1,4)$, etc. The point of our definition is that, when $\sigma \to s$, it means that σ is one of the truncations of the signed continued fraction expansion of s.

LEMMA 11.5. *If $\sigma \to s$ and $\sigma = p/q$, then $|\sigma - s| \le 6/q^2$.*

Proof: We can take the S.C.F. for s, based on the E.E. of s. We define the numbers $p_1, p_2, \ldots$ and the numbers $q_1, q_2, \ldots$ as in Equation 2.18. There is some k such that $\sigma = \sigma_k = p_k/q_k$. We have already mentioned that

$$|\sigma_{k+1} - \sigma_k| \le \frac{1}{q_k^2}. \tag{11.6}$$

A case-by-case analysis shows that

$$|q_{k+3}| \geq 2|q_k|. \tag{11.7}$$

The point here is that we have $|q_{k+2}| \geq 2|q_k|$ unless $a_{k+2} = -2$, and this cannot happen for two indices in a row.

From here it is easy to see that $\{\sigma_k\}$ is a Cauchy sequence whose limit is s. We also have the estimate

$$|s - \sigma_k| \leq \sum_{j=k}^{\infty} |\sigma_{j+1} - \sigma_j| \leq \sum_{j=k}^{\infty} \frac{1}{q_j^2} \leq \frac{6}{q_k^2}. \tag{11.8}$$

This completes the proof. □

11.6. Dense Orbits

Now we prove that almost all R-orbits are dense. It suffices to consider R-orbits of points in $(0, 1/2)$. Define $R_1 : (0, 1/2] \to (0, 1/2]$ so that $R_1(s) = R(s)$ if $R(s) < 1/2$ and otherwise $R_1(s) = 1 - R(s) = R^2(s)$.

LEMMA 11.6. *Let $s \in (0, 1/2)$. If $\{R_1^n(s)\}$ is dense in $(0, 1/2)$ then $\{R^n(s)\}$ is dense in $(0, 1)$.*

Proof: Note that $\{R^n(s)\}$ contains $\{R_1^n(s)\}$ because R_1 is always a power of R. Hence $\{R^n(s)\}$ is dense in $(0, 1/2)$. But R maps $(1/4/1/3)$ onto $(1/2, 1)$ in a continuous way. Hence $\{R^n(s)\}$ is dense in $(1/2, 1)$ as well. □

Let Γ be the $(2, 4, \infty)$ triangle group which appears in some of our results, and let T be the $(2, 4, \infty)$ hyperbolic triangle generating Γ. One of the edges of T is contained in the geodesic circle C fixed pointwise by the map $z \to 1/(2\overline{z})$. We color C red and the other two edges blue. We identify the set of (vertical) geodesics emanating from the cusp of T with $(0, 1/2]$. The recipe is that each $s \in (0, 1/2]$ corresponds to a geodesic connecting s to ∞. The identification gives a natural measure on the set of such geodesics.

Consider the family of vertical blue lines connecting ∞ to half-integers. One can describe R_1 like this. Starting with $s = s_0 \in [0, 1/2)$, we first reflect in the red circle C to produce the point s_1. There is some nearest vertical blue line which separates s_1 from 0. We reflect in this blue line to produce s_2. We now repeat these reflections in blue lines until we arrive at a point in $(0, 1/2)$, and this point is $R_1(s)$.

Now we describe a construction that is very similar to what C. Series does in [**BKS**, §5]. If γ is the vertical geodesic billiard path corresponding to s, then γ bounces off the red side and points to s_1 in the following sense. Were we to follow the new trajectory in $\boldsymbol{H}^2$ all the way to the limiting point on $\partial \boldsymbol{H}^2$, that point would be s_1. Next, γ bounces off a blue side and points to s_2, etc. When our billiard path finally hits the red side again, it points to $R(s)$. Hence, we can recover the action of R_1 on s by looking at the billiard path determined by γ. The orbit $\{R_1^n(s)\}$ is dense provided that the lift of the billiard path is dense in the unit tangent bundle of T. The following result finishes our proof.

LEMMA 11.7 (Triangle). *Almost all geodesics emanating from the cusp of T have billiard trajectories which are dense in the unit tangent bundle.*

11.7. Proof of the Triangle Lemma

LEMMA 11.8. *Almost every Gauss-orbit is dense in* $(0,1)$.

Proof: Combine Theorem 2.2, Theorem 3.18, and Exercise 5.20 in [**BKS**]. □

Let Σ_0 be the double of the ideal triangle with vertices $0, 1, \infty$. We identify Σ_0 with the quotient $\boldsymbol{H}^2/\Gamma$, where Γ is an index 6 subgroup of the modular group. The lifts of the two triangles gives rise to the famous *modular tiling* of the hyperbolic plane by ideal triangles. The ideal triangles in this tiling have rational vertices, and every rational point of $\boldsymbol{R} \cup \infty$ (including ∞) is the vertex of infinitely many ideal triangles in the tiling.

LEMMA 11.9. *Almost all geodesic rays emanating from a cusp of* Σ_0 *lift to dense subsets of the unit tangent bundle* $T_1(\Sigma_0)$.

Proof: Let β be a geodesic emanating from the cusp of Σ_0 corresponding to ∞. Let x denote the point where β limits on the real axis. We can assume that $x \in (0,1)$. Suppose x has dense image under the Gauss map. As is discussed in [**BKS**, §5.4] there is a natural correspondence between the continued fraction expansion of x and the *cutting sequence* associated to β – i.e. the pattern in which β intersects the edges of the modular tiling.

One sees every finite string of digits in the continued fraction expansion of x. Suppose we choose some pair (γ, p) where γ is a geodesic with irrational endpoints and $p \in \gamma$ is some point. We can find points $q_n \in \beta$ so that the length $2n$ segment of β centered at q_n and the length $2n$ segment of γ centered at p have the same cutting sequences. Since γ has irrational endpoints, this situation forces the sequence (β, q_n) to converge to (α, p). Hence β lifts to a dense curve in $T^1(\Sigma_0)$. □

Let Σ be a finite normal covering surface of Σ_0. That is, $\Sigma_0 = \Sigma/G$, where G is some finite group acting on Σ. There is again a natural measure on the set of geodesics emanating from a cusp of Σ.

LEMMA 11.10. *Almost all geodesic rays emanating from the cusp of* Σ *lift to dense subsets of* $T_1(\Sigma)$.

Proof: Let α be a geodesic emanating from our cusp of Σ. Let $\overline{\alpha}$ be the projection of α to Σ_0. Almost every choice of α leads to $\overline{\alpha}$ having a dense lift in $T_1(\Sigma_0)$. But then the G-orbit of the lift of α is dense in $T_1(\Sigma)$. Let C be the closure of the lift of α in $T_1(\Sigma)$. We know that $G(C) = T_1(\Sigma)$. At the same time, we know that $\overline{\alpha}$ approximates, with arbitrary precision, any closed loop in Σ_0. From this we see that in fact C is G-invariant. Hence $C = T_1(\Sigma)$, as desired. □

The $(2,4,\infty)$ triangle T is finitely covered by a normal cover Σ of Σ_0. The covering projection maps geodesics in the surface to billiard paths on Σ. Almost every billiard path on T emanating from the cusp lifts to a geodesic on Σ which emanates from the corresponding cusp and is dense in $T^1(\Sigma)$. But then the billiard path is dense in $T^1(T)$.

CHAPTER 12

Properties of the Tiling

12.1. Tedious Special Cases

To make our result work out perfectly, we have to deal with some tedious special cases which are not covered by the Main Theorem. Let $s_0 = s$ and $s_n = R^n(s)$. We list these special cases here, and also explain how we interpret the quantities which will appear in our results below:

- We define $R(1/2n) = 0$ for $n = 1, 2, 3, ...$ Thus, when s is rational, we have $s_n = 0$ for some n. This is how all the sequences end in the rational case.
- Recall that O_s is the trivial tile for all $s \in (0, 1)$. We define O_0 as the set of 4 isosceles triangles having vertices $(0, 0)$ and $(\pm 1, \pm 1)$. This case is relevant because, in the rational case, all our sequences end in 0.
- We call s *exceptional* if $s = 1/2n$ for $n = 1, 2, 3, ...$ Since $R(s) = 0$ in this case, the map ϕ_s is not defined. However, we now define

$$\phi_s(x, y) = \left(\frac{x}{2n}, \frac{y}{2n}\right), \qquad s = \frac{1}{2n}. \tag{12.1}$$

- When $s_n = 1/2$ we must have $s_{n-1} < 1/2$. We interpret $\phi_{s_{n-1}}(O_{s_n})$ as the image of the central tile under the extension of $\phi_{s_{n-1}}$ guaranteed by the Main Theorem.
- When $s < 1/2$ is exceptional, we know from the Insertion Lemma that Δ_s has at least 2 additional tiles which are translates of O_s. These additional tiles are the central tiles which lie on either side of O_s. In this case, we interpret $\phi_{s_{n-1}}(O_{s_n})$ as the image of one of these other central tiles under $\phi_{s_{n-1}}$.
- Recall that $n > 0$ is a *good index* when $s_{n-1} < 1/2$. In this case we interpret $\phi_{s_{n-1}}(O_{s_n})$ as the image of O_{s_n} under one of the two extensions of $\phi_{s_{n-1}}$, as in Statement 3 of the Main Theorem.

12.2. Classification of Tile Shapes

As we mentioned in the introduction, Theorem 1.2 describes the tiles in Δ_s up to similarity. We stated Theorem 1.2 mainly for convenience. Here we state a more precise result which immediately implies Theorem 1.2. Theorem 12.1 below describes the tile in Δ_s up to translation.

At least up to translation, the sets

$$O_s^0 = O_s, \qquad O_s^n = \phi_{s_0} \circ ... \circ \phi_{s_{n-1}}(O_{s_n}), \qquad n = 1, 2, 3, ... \tag{12.2}$$

make sense for every index n which is either good or exceptional. We use the convention that these sets are undefined the other indices.

THEOREM 12.1. *Let $s \in (0,1)$. A polygon arises in Δ_s if and only if it is a translate of O_s^n for some good or exceptional index $n \geq 0$.*

Proof of Theorem 1.2: Theorem 12.1 immediately implies Theorem 1.2 in the irrational case. In the rational case, Theorem 12.1 implies all the statements of Theorem 1.2 except possible the statement that Δ_s must contain a square tile. We will see in Lemma 12.10 below that Δ_s must have a square tile. □

To prove Theorem 12.1 we first consider the case when s is rational and then we take limits.

LEMMA 12.2. *Theorem 12.1 is true for s exceptional.*

Proof: Inspecting Figure 7.4 we see that $\Delta_{1/2}$ consists of one square and 4 right angled isosceles triangles, each one similar to one of the triangles in O_0. By the Insertion Lemma, we see that $\Delta_{1/2n}$ consists of a row of equally sized squares – the central tiles – and the 4 isosceles triangles. The maps $\phi_{1/2n}$ have been designed to get the scaling correct. □

LEMMA 12.3. *Theorem 12.1 for all rational $s \in (0,1)$.*

Proof: Let $t = R(s)$. Now we can assume that s is not exceptional. So, the Main Theorem applies to the pair (s,t). By the Main Theorem, a tile σ appears in Δ_s if and only if the following holds:

- $\sigma = O_s$.
- σ is a translate of $\phi_s(\tau)$ where $\tau \neq O_t$ is a tile of Δ_t.
- $s < 1/2$ and $\sigma = \phi_s(O_t)$.

From this menu of options, we see that the truth of Theorem 12.1 for the parameter t implies the truch of Theorem 12.1 for the parameter s. Our result now follows by induction on the length of the sequence $\{R^n(s)\}$. □

This completes the proof of Theorem 12.1 in the rational case. Now suppose s is irrational. The same argument in the rational case applies to show that some translate of each tile in Equation 12.2 does in fact appear as a tile of Δ_s. To finish the proof, we need to establish the converse.

Suppose that P is a tile of Δ_s. Let $\{{}_ns\}$ be a sequence of rationals converging to s. By Lemma 8.7, there is a tile ${}_nP$ of $\Delta_{{}_ns}$ such that ${}_nP \to P$ as $n \to \infty$. From the rational case of Theorem 12.1, there is some k_n such that ${}_nP$ is a translate of

$$ {}_nT_{k_n}(O_{u_n}), \qquad u_n = R^{k_n}({}_ns). $$

Here ${}_nT_j$, for $j = 0,1,2,...$, are the maps in Equation 12.2.

The scale factor of ${}_nT_k$ tends to 0 as $k \to \infty$, and ${}_nP$ has uniformly large diameter. Therefore, the sequence $\{k_n\}$ is a bounded sequence. Passing to a subsequence, we can assume that ${}_nk$ is independent of n. Hence ${}_nP$ is a translation of

$$ {}_nT_k(O_{u_n}), \qquad u_n = R^k({}_ns). \tag{12.3} $$

But

$$ {}_nT_k \to T_k, \qquad u_n \to u = R^k(s), \qquad O_{u_n} \to O_u. $$

Hence P is a translate of $T_k(O_u)$. This completes the proof of Theorem 12.1 in the irrational case.

Remark: The reader might wonder what happened to the triangular tiles which exist in the rational case but not in the irrational case. What is going on is that these tiles shrink to points as we pass to an irrational limit. In terms of the proof give above, we would need the sequenc $\{_nk\}$ to be unbounded in order to see these triangles in the limit, but this is not possible.

12.3. Classification of Stable Orbits

We prove Theorem 1.4 through a series of lemmas.

LEMMA 12.4. *There are 4 triangular orbits for any rational s.*

Proof: When $s = 1/2n$ there are 4 triangular tiles, each one its own orbit. The Main Theorem implies that the number of triangular orbits for s is the same as the number of triangular orbits for $t = R(s)$. So, by induction, there are always 4 triangular orbits. □

LEMMA 12.5. *There is a single period $N(s)$ such that all the unstable orbits have period $N(s)$. In particular, Δ_s contains the same number of right angled isosceles triangles in each of the 4 orientations.*

Proof: We have already established that the unstable periodic tiles group themselves into 4 orbits. Call these orbits $O_1, ..., O_4$. Rotation in the origin permutes these orbits. Let's say that rotation in the origin has the effect $O_1 \leftrightarrow O_3$ and $O_2 \leftrightarrow O_4$. The map μ_s from Calculation 1 (Lemma 9.2) also permutes the orbits, but the permutation is different on account of the fact that the linear part of μ_s is orientation reversing. The orbits can be labeled so that μ_s has the action $O_1 \leftrightarrow O_2$ and $O_3 \leftrightarrow O_4$. The existence of these permutations shows that all these orbits have the same period. □

Let $s \in (0, 1)$ be rational. Theorem 1.4 says that a periodic tile of Δ_s is stable if and only if it is not a triangle. We prove this result through a series of lemmas.

LEMMA 12.6. *Suppose that $p \in X_s$ is periodic for all s' sufficiently close to s. Then p is a stable periodic point.*

Proof: Consider the integer arithmetic graph of p, as a function of s. These graphs must be constant in a neighborhood of s, or else there would be parameters arbitrarily close to s relative to which p did not have a well-defined orbit. Since the graphs are constant in a neighborhood of s, the point p is stable by definition. □

LEMMA 12.7. *Let s and $t = R(s)$ be as in the Main Theorem. Given a periodic tile $P_t \in \Delta_t$, suppose that $\phi_s(P_t)$ is contained in the same orbit as P_s. Then P_t is stable iff P_s is stable.*

Proof: By the Insertion Lemma, we can take $s \in [1/4, 1)$. Note that R is not defined on $s = 1/4$ and $s = 1/2$. So, either $s \in (1/4, 1/2)$ or $s \in (1/2, 1)$. In either situation, the same case of the Main Theorem holds for all parameters sufficiently near s, and the map R is continuous in a neighborhood of s.

Suppose P_t is stable. Any $p \in P_t$ is stable. Let $q_{s'} = \phi_{s'}(p)$. By the Main Theorem, $q_{s'}$ is periodic for all s' sufficiently close to s. Hence q_s is stable. Hence, the periodic tile $\phi_s(P_t)$ containing q_s is stable. But then P_s is stable.

The converse has a similar proof. □

LEMMA 12.8. *A tile of Δ_s is unstable if and only if it is a triangle.*

Proof: We check the truth of the result for $s = 1/2$. The case $s = 1/(2n)$ follows from the Insertion Lemma. In general, the result follows from Lemma 12.7 and induction on the number n such that $R^n(s) = 0$. The one case we need to worry about is when $s < 1/2$ and $t = R(s)$. In this case, the only tiles of Δ_s not directly covered by Lemma 12.7 are the tiles of order 2 coming from the image of the trivial tile under the extensions of ϕ_s. But the Main Theorem applies here as well, for such tiles are the images of the trivial tile under the extensions of the two halves of ϕ_s. □

These lemmas together complete the proof of Theorem 1.4.

12.4. Existence of Square Tiles

In this section we prove Statement 1 of Theorem 1.3.

We mention first of all that when $s = \sqrt{2}/2$, there are no square tiles at all. See Figure 1.7. In this case $R^n(s) = \sqrt{2}/2$ when n is even and $R^n(s) = 1 - s$ when n is odd. In this case, n is a good index if and only if n is even. Hence, the tiles O_s^n are defined when n is even, and they are all regular octagons. Thus Δ_s consists entirely of regular octagons. This is one of the main points of the paper [**AKT**], and the proof there also goes by way of renormalization.

LEMMA 12.9. *If $s \neq \sqrt{2}/2$ is irrational, then Δ_s has a square tile.*

Proof: Suppose first that the even expansion of s is not entirely odd. Then, by Lemma 11.2, there is some k such that $R^k(s) < 1/2$ and $R^{k+1}(s) < 1/2$. But then, in the notation of Theorem 12.1, there is some n such that n is good and O_{s_n} is a square. Theorem 12.1 finishes the proof in this case.

Suppose that the even expansion of s is entirely odd. We set $\Delta_n = \Delta_{s_n}$ for ease of notation. Let $\{e_n\}$ be the even expansion of s. If $e_n < 5$ for all n, then the even expansion must be either $(3, 1, 3, 1, ...)$ or $(3, 1, 3, 1, ...)$. In the first case, $s < 1/2$ and the trivial tile is a square. In the second case $s = \sqrt{2}/2$.

The only case left is when there is some n such that $e_n \in \{5, 7, 9, ...\}$. But then, by the Insertion Lemma, some central but nontrivial tile of Δ_n is a square. But then some noncentral tile of Δ_{n-1} is a square. Repeated applications of the Main Theorem now show that some tile of Δ_0 is a square. □

LEMMA 12.10. *If $s \neq \sqrt{2}/2$ is rational, then Δ_s has a square tile.*

Proof: The argument is the same as the irrational case unless the even expansion of s has the form $(1, 3, 1, 3, ..., k)$, where just the last term k is even. If $k = 2$ then $s_n = 1/2$ and $s_{n-1} < 1/2$. Here $\Delta_{s_{n-1}}$ contains a noncentral square tile, the image of the central square $O_{1/2}$ under the extension of the map $\phi_{s_{n-1}}$ from the Main Theorem. If $k = 4, 6, 8, ...$ then n is an exceptional index and a similar argument shows that $\Delta_{s_{n-1}}$ contains a noncentral square tile. But now the same inductive argument as in the irrational case shows that Δ_s contains a noncentral square tile. □

LEMMA 12.11. *If Δ_s contains finitely many squares, then $s \in \boldsymbol{Q}[\sqrt{2}]$.*

Proof: Suppose Δ_s has finitely many squares. Then there is some $u = R^n(t)$ such that Δ_u has no squares. But then $R^n(t) = \sqrt{2}/2$. This implies that $t \in \boldsymbol{Q}[\sqrt{2}]$. □

The lemmas in this section combine to prove Statement 1 of Theorem 1.3.

Remark: Lemma 12.11 is somewhat imprecise. A more precise statement would be that Δ_s contains finitely many squares if and only if the even expansion of s eventually agrees with that of $\sqrt{2}/2$. That means, it terminates as $1, 3, 1, 3, ...$

12.5. The Oddly Even Case

Statement 2 of Theorem 1.3 follows from what we have already said. Suppose s is irrational and the continued fraction of s has the form $(0, s_1, s_2, ...)$ with s_k even for all odd k. By the characterization given in §11.3, we know that $R^n(s) < 1/2$ for all n. But then every tile of Δ_s is a square by Theorem 12.1.

12.6. Density of Shapes

Statement 3 of Theorem 1.3 also follows from what we have already said. By Theorem 12.1, the set Δ_s contains a dense set of shapes of semi-regular octagons, and an infinite number of squares, provided that the orbit $R^n(s)$ is dense in $(0, 1)$. But, since almost every R-orbit is dense, this holds.

Part 3

Metric Properties

Here is an overview of this part of the monograph:

- In §13 we elaborate on how the Main Theorem relates the tiling Δ_s to the tiling Δ_t for $t = R(s)$. This rather technical chapter is important for understanding the global structure of the tiling. We call our main result the Filling Lemma.
- In §14 we give a second elaboration of the Main Theorem, this time establishing a kind of recursive decomposition of the tiling Δ_s into the similar copies of the pieces from §8. We call this result the Covering Lemma.
- In §15 we prove several auxilliary geometric facts about Δ_s.
- In §16 we put together the work from §13-15 and prove Theorems 1.5 and 1.10.
- In §17 we prove Theorems 1.9 and a related result, Corollary 17.9.
- In §18 we give a recursive formula for the the period $N(p/q)$ of the unstable orbits of $f_{p/q}$.
- In §19 we deduce Theorem 1.6 from Corollary 17.9 and the formula in §18.

CHAPTER 13

The Filling Lemma

13.1. The Layering Constant

We were able to get a lot of information about the tiling Δ_s just from the Main Theorem and symmetry. However, to prove more subtle results, such as the fact that Δ_s has full measure, we need more information about the Main Theorem. The basic limitation of the Main Theorem is that the subset

$$\phi_s(\Delta_t \cap Y_t) = Z_s \cap \Delta_s \tag{13.1}$$

could be a very small subset of Δ_s. Looking at pictures such as Figure 9.3, we see that many copies of the set in Equation 13.1 can be layered on top of each other. In this chapter, we elaborate on this structure.

The constant we introduce in this section essentially determines how many copies of $\phi_s(\Delta_t \cap Y_t)$ we see in Δ_s. When $s < 1/2$ we let $t = R(s)$ and $u = R(t)$.

- If $s < 1/2$ and $t > 1/2$ let $\mho = 1$.
- If $s < 1/2$ and $t < 1/2$ and $u < 1/2$, let $\mho = \text{floor}(1/(2t))$.
- If $s < 1/2$ and $t < 1/2$ and $u > 1/2$, let $\mho = 1 + \text{floor}(1/(2t))$.
- If $s > 1/2$ let

$$\mho = \text{floor}\left(\frac{1}{2-2s}\right)$$

We call $\mho$ the layering constant.

Here we explain the value of $\mho$ in terms of the even expansion. We write $\mho(a,b) = k$ if $\mho(s) = k$ whenever the even expansion of s starts out $(a, b, ...)$. We have

- $\mho(a,1) = 1$ if $a > 1$ is odd.
- $\mho(a,b) = b/2$ if a and b are even.
- $\mho(a,b) = (b+1)/2$ when a is even and $b > 1$ is odd.
- $\mho(1,b) = b/2$ when b is even.
- $\mho(1,b) = (b-1)/2$ when b is odd.

13.2. The Filling Lemma, Part 1

We first describe our result for the case $s < 1/2$. Our discussion refers (initially) to Figure 13.1 below, which shows the relevant sets for two different parameters. Let U_s be the image, under of ϕ_s, of the trivial tile in Δ_t. These are is the dark red tiles in Figure 13.1. Define

$$\Psi_s^0 = Z_s^0 \cup U_s. \tag{13.2}$$

Ψ_s^0 is the (light and dark) red set on each half of Figure 13.1. Let τ_s denote the subset of Z_s^0 lying beneath the line extending the top right edge of U_s. Here τ_s is the colored region in each half of Figure 13.1.

Figure 13.1: The relevant sets for parameters 13/44 and 11/26.

Let T_s denote the transformation which translates by a vector pointing in the positive x direction and having length equal to the length of the bottom side of Ψ_s^0. We have a partition

$$\tau_s = \bigcup_{j=0}^{\mho(s)} \Psi_s^j, \qquad \text{where} \qquad \Psi_s^j = T_s^j(\Psi_s^0) \cap \tau_s \tag{13.3}$$

Here $\mho(s)$ is the layering constant. For larger j, the sets in Equation 13.3 are empty.

The whole tiling is determined by the tiling inside τ_s and symmetry. Let P_s, and its symmetry R_D, be as in §9. In terms of sets, we have

$$X_s = X_s^0 \cup \text{central tiles} \cup \iota(X_s^0), \qquad X_s^0 = \tau_s \cup R_D(\tau_s \cap P_s). \tag{13.4}$$

The second equation is a consequence of the fact that the top right boundary of τ_s is parallel to the line of symmetry D_s of R_D, and above it.

The following result explains the structure of Δ_s inside τ_s. What the result says is that the translaton T_s^{-j} respects the tiling of Δ_s on the relevant domain.

LEMMA 13.1 (Filling). *$T_s^{-j}(\Delta_s \cap \Psi_s^j) = \Delta_s \cap T^{-j}(\Psi_s^j)$ for all $j = 1, ..., \mho$.*

Proof: We remind the reader that our system is given by $f_s : X_s \to X_s$.

By the Insertion Lemma, we can take $s \in (1/4, 1/2)$. The map f_s is defined in terms of a partition of X_s. On each piece of this partition, f_s is a translation. Let Ω_s be the piece of the partition which shares the lower left vertex of X_s^0. A routine calculation shows that the restriction of f_s to Ω_s is T_s.

When $s \in (1/3, 1/2)$, the set Ω_s is the quadrilateral with the following properties.

- The left edge of Ω_s is contained in the left side of X_s.
- The bottom edge of Ω_s is contained in the bottom edge of X_s.
- The top edge of Ω_s lies in the same line as the top edge of Z_s^0.
- The right edge e_s of Ω_s is vertical and has the property that $T_s(e)$ lies in the left edge of the leftmost central tile of Δ_s.

When $s \in (1/4, 1/3)$, the set Ω_s is a triangle whose left, right, and bottom sides are as above. Our argument works the same in either case.

Figures 13.2, 13.3, 13.4 below show the picture for $s = 19/60, 11/30, 9/20$. The diagonal line L_s bisects the yellow square and the green diamond in each picture.

A routine calculation shows that

$$T_s^{-1}(\tau_s - \Psi_s^0) \subset \Omega_s. \tag{13.5}$$

If we start with any point $p \in \tau_s - \Psi_s^0 = Z_s^0 \cup U_s$, we see from the shape of Ω_s^0 that the iterates $f_s^{-j}(p)$ are defined and lie in Ω_s^0, for each $j = 0, 1, 2, ...$ until we reach some $k \leq \mho$ such that $f_s^{-k}(p) \subset Z_s^0$. Our result follows from this observation and from the fact that Δ_s is f_s-invariant. $\square$

Figure 13.2: Z_s^0 (red, orange), Ω_s (pink) and $f_s(\Omega_s)$ green

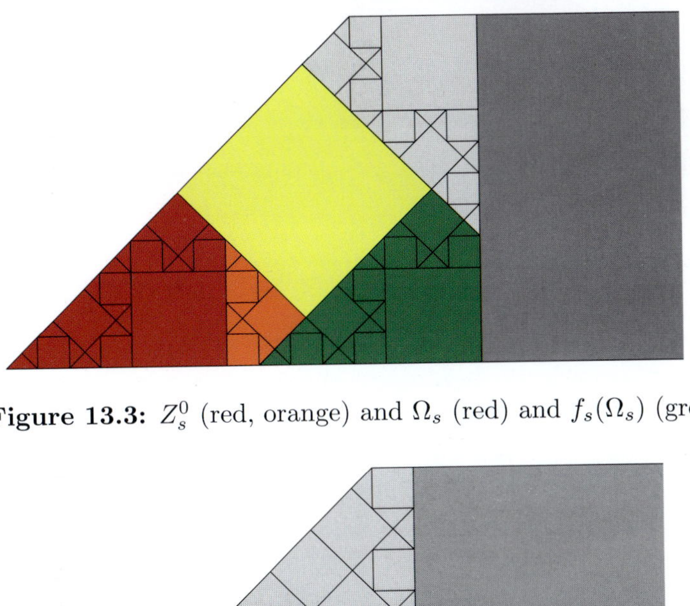

Figure 13.3: Z_s^0 (red, orange) and Ω_s (red) and $f_s(\Omega_s)$ (green).

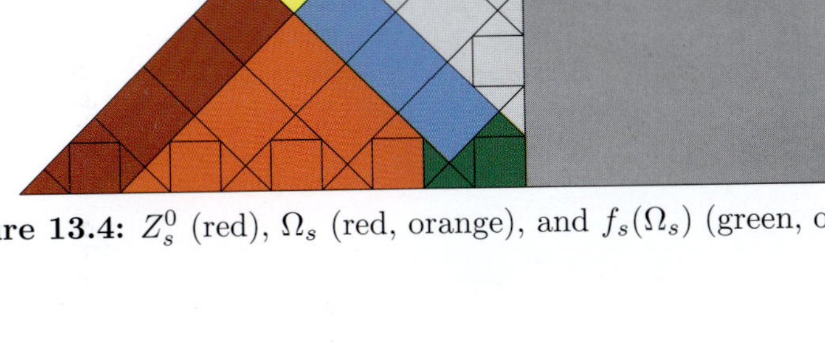

Figure 13.4: Z_s^0 (red), Ω_s (red, orange), and $f_s(\Omega_s)$ (green, orange, blue).

13.3. The Filling Lemma, Part 2

Now we explain the picture for $s \in (1/2, 1)$.

This time, the right edge of Z_s^0 lies on the same line as the left edge of the central tile of Δ_s. Let δ_s denote this line. Let T_s denote the translation by the vector which is positivel proportional to $(1,1)$ and whose length is the same as the length of the left side of Z_s^0. Let τ_s be the region of X_s^0 lying to the left of δ_s. In Figure 13.5, the region τ_s is colored.

Figure 13.5: Ψ_s^j for $j = 1, 2, 3$ (red, blue, yellow) for $s = 14/17$.

We have a partition

$$(13.6) \qquad \tau_s = \bigcup_{j=0}^{\upsilon}(s)\Psi_s^j, \qquad \text{where} \qquad \Psi_s^j = T_s^j(Z_s) \cap \tau_s.$$

The whole tiling is determined by the tiling inside τ_s and symmetry. Let A_s, and its symmetry R_V, be as in §9. In terms of sets, we have

$$(13.7) \qquad X_s = X_s^0 \cup \text{central tiles} \cup \iota(X_s^0), \qquad X_s^0 \subset \tau_s \cup R_V(\tau_s \cap A_s).$$

With these definitions, the Filling Lemma holds *verbatim*, and the proof is essentially the same. This time Ω_s is a right isosceles triangle, and the left edge of $f_s(\Omega_s)$ lies in δ_s, and f_s translates diagonally along the vector that generates the left side of Z_s. In Figure 13.5, Ω_s is the union of the light red and light blue tiles.

CHAPTER 14

The Covering Lemma

14.1. The Main Result

In this chapter we establish a variant of the Filling Lemma. The variant gives a more precise recursive description of the set Δ_s.

In §9 we defined 4 bilaterally symmetric subsets $A_s, B_s, P_s, Q_s \subset X_s^0$. We call these sets the *symmetric pieces*. An *ϵ-patch* is a triple (K, ψ, u) where

- $u \in (0,1)$.
- K is one of the 4 symmetric pieces A_u, B_u, P_u, Q_u.
- $\psi : K \to X_s$ is a similarity which contracts by some factor $\lambda \leq \epsilon$.
- $\psi(\Delta_u \cap K) = \Delta_s \cap \psi(K)$.

The last condition means that ψ gives a bijection between tiles of $\Delta_u \cap K$ and tiles of $\Delta_s \cap \Psi(K)$. When we have an ϵ-patch (K, ψ, u), we are recognizing a small portion of Δ_s as being a similar copy of a large portion of Δ_u. When the choice of ϵ is not relevant to the discussion, we will just say *patch*.

LEMMA 14.1 (Covering). *For any $\epsilon > 0$, each symmetric piece is partitioned into a finite union of tiles and a finite union of ϵ-patches.*

Lemma 14.1 comes from iterating Lemma 14.2.

LEMMA 14.2. *Suppose $s \in (0,1)$ is irrational, and let $t = R(s)$. Each symmetric piece is partitioned into a finite union of tiles and a finite union of patches of the form (K_t, ϕ_s, t).*

The rest of the chapter is devoted to proving Lemma 14.2. We will do it in a case-by-case manner. The cases depend mostly on the layering constant associated to s.

Let $\mho = \mho(s)$ be the layering constant for s which appears in the Filling Lemma. Let R_D, R_V, etc. be the symmetries discussed in §9. We use the notation $X \sim Y$ to denote that X and Y agree up to the insertion or deletion of a finite union of tiles. We define

$$\nu(a,b) = \bigcup_{a}^{b} \Psi_t^k. \tag{14.1}$$

As long as $k < \mho(t)$, this set is a union of patches of the form discussed in Lemma 14.2. For instance $\Psi_t^0 \sim \phi_s(A_t) \cup \phi_s(B_t)$.

LEMMA 14.3. *Lemma 14.2 is true when* $s \in (3/4, 1)$.

Proof: In this case $\mho(s) \geq 2$. Figure 14.1 shows a picture.

Figure 14.1: Δ_s^0 for $s = 4/5$ and Δ_t^0 for $t = R(s) = 1/5$. Here $\mho(s) = 2$.

(1) A_s is green.

$$A_s = R_D \phi_s(B_t).$$

Here $\phi_s(B_t)$ is dark red.

(2) B_s is red/pink/yellow/blue.

$$B_s \sim Q_s \cup R_V(Q_s) \cup R_D \phi_s(A_t).$$

Here Q_s is blue and $\phi_s(A_t)$ is light red. This equation reduces the case of B_s to the case of Q_s.

(3) P_s is red/pink/yellow/green.

$$P_s = \nu(0, \mho - 1) \cup R_D \phi_s(Q_t).$$

Here $R_D \phi_s(Q_t)$ is light green.

(4) Q_s is blue.

$$Q_s \sim R_V(\nu) \cup R_E R_D \phi_s(Q_t), \qquad \nu = \nu(0, \mho - 2).$$

This completes the proof. $\square$

LEMMA 14.4. *Lemma* 14.2 *is true when* $s \in (1/2, 3/4)$.

Proof: In this case $\mho(s) = 1$.

Figure 14.2: Δ_s^0 for $s = 9/14$ and Δ_t^0 for $R(s) = 5/14$. Here $\mho(s) = 1$.

Here are the decompositions in this case.

(1) A_s is yellow/green. $A_s = R_D \phi_s(B_t)$. (As in the previous case.) $\phi_s(B_t)$ is (dark red)/pink.

(2) B_s is red/pink/blue.

$$B_s = R_d \phi_s(A_t) \cup \phi_s(Q_t) \cup R_V \phi_s(Q_t).$$

The first set on the right is (light red)/pink, the second set is dark red, and the third set is blue.

(3) P_s is red/pink/green/yellow.

$$P_s = \phi_s(P_t) \cup \phi_s(Q_t) \cup R_D \phi_s(Q_t).$$

The first set on the right is pink/red/green, and the second set is dark red, and the third set is yellow.

(4) Q_s is blue. $Q_s = R_V \phi_s(Q_t)$.

This completes the proof. □

LEMMA 14.5. *Lemma 14.2 is true when* $s \in (1/3, 1/2)$.

Proof: In this case $R(s) < 1/2$. Let $\mho' = \mho - 1$ when $R(t) < 1/2$ and $\mho' = \mho - 2$ when $R(t) > 1/2$.

Figure 14.3: Δ_s^0 for $s = 7/17$ and Δ_t^0 for $t = R(s) = 3/14$. Here $\mho(s) = 2$.

(1) A_s is blue/yellow/red.

$$A_s \sim R_H R_D(E_s) \cup R_D(E_s) \cup R_D(\nu') \cup R_D R_V \phi_s(Q_t).$$

$$E_s = \phi_s(A_t) \qquad \nu' = \nu(1, \mho').$$

A_s is blue/red/yellow. The first set on the right is yellow. The second set is red. The union of the third and fourth sets is dark blue. ν is medium green.

(2) B_s is green.

$$B_s = \nu \cup R_V \phi_s(Q_t), \qquad \nu = \nu(0, \mho').$$

The set $R_V \phi_s(Q_t)$ is light green, just beneath the yellow tiles.

(3) P_s is everything colored except the dark green corner. $P_s = A_s \cup R_D(A_s)$.

(4) Q_s is the dark green corner. $Q_s = \phi_s(B_t)$.

This completes the proof. □

LEMMA 14.6. *Lemma* 14.2 *is true when* $s \in (1/4, 1/3)$.

Proof: In this case $R(s) > 1/2$.

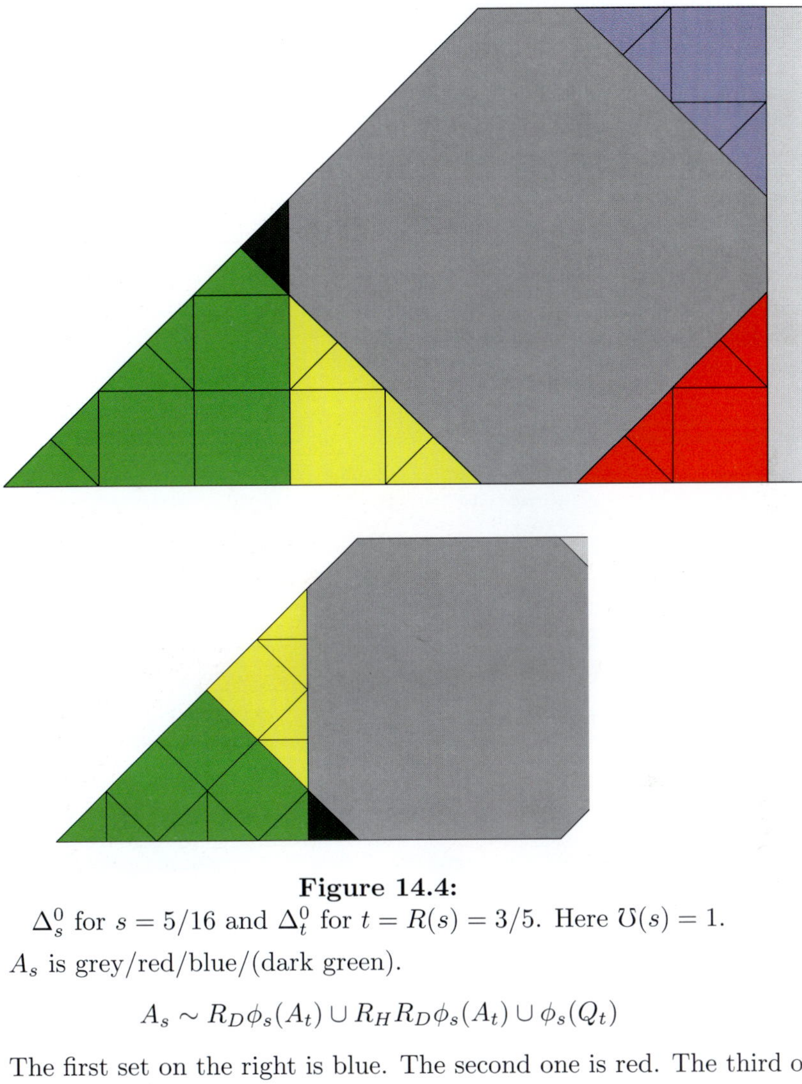

Figure 14.4:
Δ_s^0 for $s = 5/16$ and Δ_t^0 for $t = R(s) = 3/5$. Here $\mho(s) = 1$.

(1) A_s is grey/red/blue/(dark green).

$$A_s \sim R_D\phi_s(A_t) \cup R_H R_D\phi_s(A_t) \cup \phi_s(Q_t)$$

The first set on the right is blue. The second one is red. The third one is dark green.

(2) B_s is yellow/(light green). $B_s = \phi_s(P_t)$.

(3) P_s is grey/red/yellow/blue.

$$P_s \sim \phi_s(A_t) \cup R_D\phi_s(A_t) \cup R_H R_D\phi_s(A_t).$$

The first set on the right is yellow. The second one is blue. The third one is red.

(4) Q_s is green. $Q_s = \phi_s(B_t)$.

This completes the proof. □

14.2. Some Additional Pictures

The decompositions in the lemmas above are meant for all parameters in the given range, even though we are illustrating the proofs with single representative pictures. Ideally, we would show all possible pictures, but this is impossible in a printed work. Here we show some additional pictures in each range. The reader can see much more using our program, OctaPet.

Figure 14.5: Δ_s^0 for $s = 7/9, 6/7 \in (3/4, 1)$.

Figure 14.6: Δ_s^0 for $s = 4/7, 5/7 \in (1/2, 3/4)$.

Figure 14.7: Δ_s^0 for $s = 5/13, 3/7 \in (1/3, 1/2)$.

Figure 14.8: Δ_s^0 for $s = 2/7, 3/10 \in (1/4, 1/3)$.

CHAPTER 15

Further Geometric Results

15.1. The Area Lemma

For each symmetric piece $K \in \{A, B, P, Q\}$ and each parameter s, define

$$\lambda(K_s) = \frac{\text{area}(K_s \cap \Delta_s)}{\text{area}(K_s)}. \tag{15.1}$$

LEMMA 15.1 (Area). *The function $\lambda(K, s)$ is uniformly bounded away from 0, for all $K \in \{A_s, B_s, P_s, Q_s\}$ and $s \in (0, 1/2)$.*

Proof: It suffices to take s to be irrational. Let $t = R(s)$. Referring to the Covering Lemma, we call K_s *active* if K_s divides into more than one patch of the form L_t, and otherwise *passive*.

Consider a passive patch, say A_s when $s \in (1/2, 1)$. We have $A_s = R_D\phi_s(B_t)$. Since the similarity $R_D \circ \phi$ respects the tilings, we have $\lambda(A, s) = \lambda(B, t)$. The same goes for the other passive patches. Inspecting the proof of the Covering Lemma, we see that a passive patch eventually subdivides into an active patch. To explain what we mean, we continue with our example. A_s subdivides into B_t, where $t = 1 - s$. By the Insertion Lemma, we can find a new parameter $t' \in (1/4, 1/2)$ such that $B_{t'}$ and B_t are similar. This leads us to consider B_s for $s \in (1/4, 1/2)$. The piece B_s is passive when $s \in (1/4, 1/3)$ and active when $s \in (1/3, 1/2)$. In the former case $B_s = \phi_s(P_t)$, and P_t is active. The several other cases have similar treatments. So, it suffices to prove our result for the active patches.

If K_s is an active patch, then we have

$$K_s = \Theta \cup \bigcup_{j=1}^{O(\mho)} K'_t. \tag{15.2}$$

Here Θ is a finite union of tiles and the union is taken over the patches given by the relevant case of the Covering Lemma. The notation $O(\mho)$ indicates that there are on the order of $\mho$ patches in the union. Here $\mho$ is the layering constant. As long as $\mho$ is uniformly bounded, the diameter of the smallest tile in Θ is bounded away from 0. So, we just have to worry about the case when $\mho$ is large. In all cases, the diameter of each patch K'_t is $O(\mho^{-1})$, and so each such patch takes up area $O(\mho^{-2})$. But then $\lambda(K_s) \to 1$ as $\mho(s) \to \infty$. Figures 15.2 and 15.3. illustrate what is going on. $\square$

15.2. Tiles in Symmetric Pieces

The goal of this section is to prove the following result:

LEMMA 15.2. *Let $s \in (0,1)$ be irrational. For each symmetric piece K_s and each edge e_s of K_s, there is a periodic tile of Δ_s having an edge in e.*

We prove Lemma 15.2 through a series of smaller results.

LEMMA 15.3. *Let $t = R(s)$. If Lemma 15.2 holds true for t then Lemma 15.2 also holds true for s.*

Proof: By the Covering Lemma, every edge of K_s contains an edge of a similar copy of some symmetric piece K'_t. By this we mean that there is a patch (K', ψ, t) so that $\psi(K') \subset K$ and one edge of $\psi(K')$ is contained in the relevant edge of K. Thus, K_s inherits the desired tile from K'_t. □

In view of the previous result, it suffices to prove Lemma 15.2 for the case $s < 1/2$.

LEMMA 15.4. *Lemma 15.2 is true for B_s and Q_s.*

Proof: Suppose $K_s \in \{B_s, Q_s\}$. Applying the Main Theorem repeatedly, we see that there are infinitely many tiles which have edges in the bottom edge of X_s and infinitely many tiles which have edges in the left edge of X_s. Eventually these tiles lie in both B_s and Q_s. This takes care of two out of three edges of B_s and Q_s. The result for the third edge, in each case, follows from bilateral symmetry. □

LEMMA 15.5. *Lemma 15.2 is true when $s < 1/2$ and $R(s) > 1/2$.*

Proof: In this case, by the Main Theorem, the left branch of ϕ_s extends to the central tile O_t, and $\phi_s(O_t)$ has edges in all 5 sides of A_s and P_s. See Figure 15.1. We have already taken care of B_s and Q_s above. □

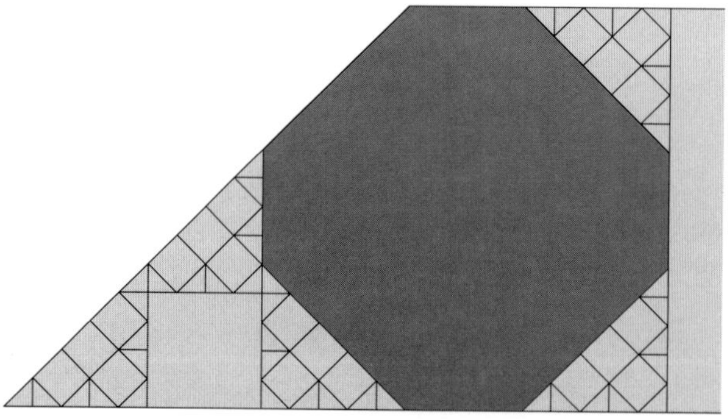

Figure 15.1: The big octagon for $s = 7/23$.

LEMMA 15.6. *Lemma 15.2 is true provided $R^n(s) > 1/2$ for some n.*

Proof: Combine Lemma 15.3 and the previous result. □

LEMMA 15.7. *Lemma* 15.2 *is true when* $R(s), R^2(s) < 1/2$.

Proof: Let $t = R(s)$ and $u = R(t)$. Let τ_s denote the central tile of Δ_s, and similarly for t and t. The tiles τ_t and τ_u are both squares, and various of the squares

$$\phi_s(\tau_t), \qquad \phi_s \circ \phi_t(\tau_u), \qquad R_D \circ \phi_s \circ \phi_t(\tau_u), \qquad R_V \circ \phi_s \circ \phi_t(\tau_u) \tag{15.3}$$

work for all the edges of A_s and P_s. Here R_D and R_V are reflections in the diagonal and vertical lines of symmetry. □

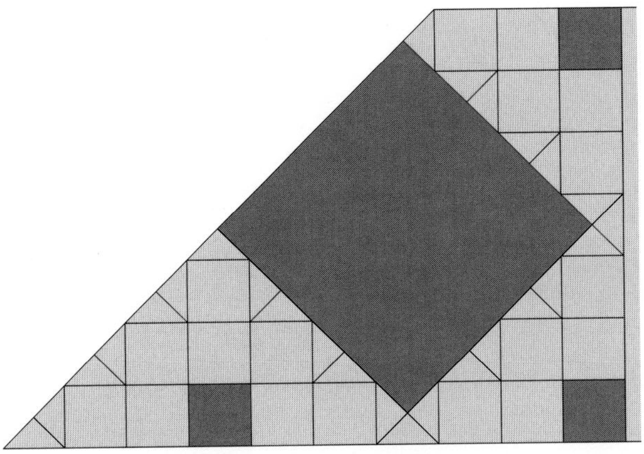

Figure 15.2: The 4 square tiles for $s = 7/20$.

We have exhausted the cases. This completes the proof of Lemma 15.2.

15.3. Pyramids

Say that a *pyramid* of size k is a configuration of squares having the structure indicated in Figure 15.3 for $k = 1, 2, 3$.

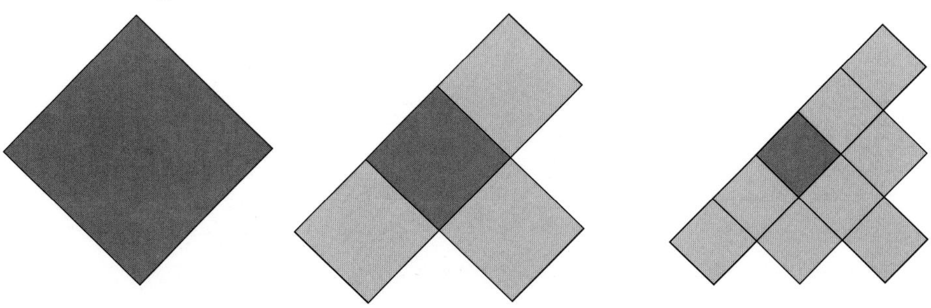

Figure 15.3: Pyramids for $k = 1, 2, 3$.

We will suppose that $s < 1/2$ and $t = R(s) < 1/2$. In this case Δ_s^0 has a square tile Q_s whose left vertex is the left vertex of the horizontal line H of symmetry. We define Ψ_s to be the union of all the tiles in Δ_s^0 which are translation equivalent to Q_s. These are the largest noncentral tiles in Δ_s. Figure 15.4 shows an example.

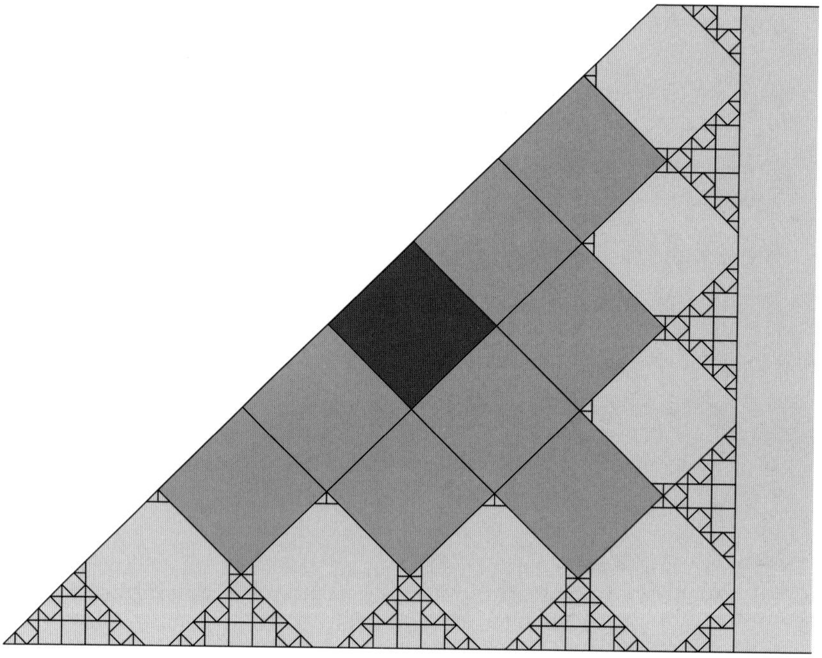

Figure 15.4: Q_s (dark) and Ψ_s (shaded) for $s = 27/61$.

LEMMA 15.8. *Let $\mho = \mho(s)$ be the layering constant for s. Then Ψ_s is a pyramid of size $\mho - 1$.*

Proof: The bottom $\mho - 1$ squares in the base of the desired pyramid are guaranteed by the Main Theorem and the Insertion Lemma (applied to Δ_t). The bottom half of the pyramid is then guaranteed by the Filling Lemma. The top half is then guaranteed by the bilateral symmetry corresponding to the diagonal line D_s. □

CHAPTER 16

Properties of the Limit Set

16.1. Elementary Topological Properties

Here we establish the result that the periodic orbits are dense. This is a weak precursor to our results about area and dimension.

LEMMA 16.1. *Δ_s is dense in X_s for all s.*

Proof: Lemma 3.2 takes care of the rational case. Suppose that s is irrational. Let $p \in X_s$ be any point. Given any $\epsilon > 0$ the Covering Lemma implies that we can find an ϵ-patch (K, ψ, u) such that $p \in \psi(K)$. But ψ maps $\Delta_u \cap K$ to $\Delta_s \cap \Psi(K)$. Now, $\Delta_u \cap K$ certainly contains a periodic tile, by Lemma 15.2. But then $\Delta_s \cap \psi(K)$ also contains a periodic tile, and this tile lies within ϵ of p. □

It now follows from Lemma 2.6 that Λ_s consists of those points p such that every neighborhood of p contains infinitely many tiles of Δ_s.

LEMMA 16.2. *When s is irrational, Λ_s has no isolated point.*

Proof: Suppose, for the sake of contradiction, that then there is some $p \in \Lambda_s$ and some open disk U containing p such that $U \cap \Lambda_s = p$. The open set U must contain infinitely many tiles of Δ_s, because $\Lambda_s \cap U$ is nonempty. Therefore, if we write Δ_s as in the Covering Lemma, the image of some patch must have p as an accumulation point. Choosing ϵ small enough, we can guarantee that there exists an ϵ-patch (K, ψ, ϵ) such that $\psi(K) \subset U$.

If K is a triangle, then two of the vertices v_1 and v_2 of K have acute angles. (The angle is $\pi/4$.) These vertices must be accumulation points of infinitely many tiles, because all the tiles are squares and semi-regular octagons. But then $\psi(v_1)$ and $\psi(v_2)$ are accumulation points of infinitely many squares in Δ_s. Hence $\Lambda_s \cap U$ contains at least 2 points. This is a contradiction.

If K is a pentagon, then K has 2 vertices v_1 and v_2 with obtuse angles. (The angle is $3\pi/4$.) If v_1 is not an accumulation point of infinitely many tiles of $\Delta_u \cap K$, then v_1 is the vertex of some octagon of $\Delta_u \cap K$, But then there are two new acute vertices w_1 and w_2 which must be accumulation points of infintely many tiles of $\Delta_u \cap K$. This gives us the same contradiction as above. The only way out of the contradiction is for both v_1 and v_2 to be accumulation points of infinitely many tiles of $\Delta_u \cap K$, but this is again a contradiction. There is no way out. □

16.2. Zero Area

Now we turn to the proof that Δ_s has full measure, or equivalently that Λ_s has zero area. This is Statement 1 of Theorem 1.5. We first recall a basic result from measure theory.

Say that a *K-disk* is a compact set D which is contained in a disk of radius Kr and contains a disk of radius r, for some r. For example, the symmetric pieces A_s, B_s, P_s, Q_s are, say, 10-disks for every s. The constant 10 is a convenient but fairly arbitrary choice.

Here is a special case of the well-known Lebesgue Differentiation Theorem. A proof can be found, e.g., in G. B. Folland's graduate textbook on real analysis.

LEMMA 16.3. *Fix K. Suppose that $S \subset \mathbf{R}^2$ is a bounded measurable set of positive Lebesgue measure. Then almost every point $p \in S$ has the following property. If $\{D_n\}$ is a sequence of K-disks containing p, having diameter shrinking to 0, then $\mu(D_n \cap S)/\mu(S) \to 1$, as $n \to \infty$.*

The points $p \in S$ satisfying the conclusion of Lemma 16.3 are called *Lebesgue points*.

Now we turn to Statement 1 of Theorem 1.5. We will argue by contradiction. Suppose that Λ_s has positive measure. Let $p \in \Lambda_s$ be a Lebesgue point. Then, by the Covering Lemma, we can find a $1/n$-patch (K_n, ψ_n, u_n) so that $p \in \psi_n(K_n)$. Let λ be the function defined in connection with the Area Lemma in §15.1. We have

$$\lim_{n\to\infty} \lambda(K_n, u_n) = \lim_{n\to\infty} \lambda(\psi_n(K_n)) = 0. \tag{16.1}$$

The first equality, which actually holds for each n, comes from the definition of a patch. The second equality comes our characterization of Λ, and also from the fact that p is a Lebesgue point. But Equation 16.1 contradicts the Area Lemma for large n. This contradiction shows that Λ_s has area 0.

COROLLARY 16.4. Almost every point of the system (X_s, f_s) is periodic.

This corollary is yet another way to see that Δ_s is dense in X_s.

16.3. Projections of the Limit Set

Now we prove Statement 2 of Theorem 1.5.

LEMMA 16.5. *Let π be projection onto some line parallel to an 8th root of unity. Then $\pi(\Lambda_s)$ contains a line segment when s is irrational.*

Proof: We first assume that π is projection onto a horizontal line. Since Λ is a closed set, it suffices to prove that $\pi(\Lambda)$ is dense in a segment.

We ignore the countably many vertical lines containing vertices of tiles in Δ. Let L be an otherwise arbitrary vertical line which contains both a point on the bottom edge b of X and a point on the left edge ℓ of X. We will prove that L contains a point of Λ; our projection result follows immediately.

Suppose L does not intersect Λ. Then L only intersects finitely many tiles, $\tau_1, ..., \tau_n$. We order these tiles according to when L enters them as we move upwards along L. Note that L must leave τ_i and enter τ_{i+1} at the same point. Otherwise, the segment of L lying between these two tiles would be the accumulation point of infinitely many tiles. Here we are using the density of Δ. For the same reason, the bottom edge of τ_1 must lie in b and the top edge of τ_n must lie in ℓ.

Since L leaves τ_i and enters τ_{i+1} at the same point, τ_i and τ_{i+1} must share an edge, and this shared edge contains a point of L. In particular, if L leaves τ_i through a horizontal edge, then L enters τ_{i+1} through a horizontal edge. Since L is a vertical line, and the tiles of Δ are semi-regular octagons with sides parallel to the 8th roots of unity, L enters τ_i through a horizontal edge if and only if L leaves τ_i through a horizontal edge.

We know that L enters τ_1 through a horizontal edge. Using the properties above, we see that L leaves τ_n through a horizontal edge. But the top edge of τ_n, which is contained in ℓ, is not horizontal. This is a contradiction. Hence L does intersect Λ.

Suppose we project onto a line parallel to a different 8th root of unity, ω. Unless $\omega = \pm i$, we can find lines perpendicular to ω which intersect both a horizontal and a diagonal edge of X. Once we have such lines, the argument we gave for $\omega = \pm 1$ works in this new context.

We just have to worry about the case $\omega = \pm i$. In this case, we observe that some edge of the trivial tile in Δ is vertical. Thus, there are horizontal lines connecting ℓ to this vertical edge. Now we run the same argument again, interchanging the roles played by the horizontal and vertical directions. □

COROLLARY 16.6. When s is irrational, $\dim(\Lambda_s) \geq 1$.

Proof: It follows immediately from the definition of Hausdorff dimension that $\dim(S_1) \geq \dim(S_2)$ provided that there is a distance non-increasing map from S_1 onto S_2. In particular, $\dim(S_1) \geq 1$ if S_1 projects onto a line segment S_2. Therefore $\dim(\Lambda_s) \geq 1$ when s is irrational. □

Here is a stronger result. Let Λ_s^* denote the set of points of Λ_s contained in the interior of X_s.

COROLLARY 16.7. When s is irrational, the projection of Λ_s^* onto a vertical line contains a line segment. In particular, $\dim(\Lambda_s^*) \geq 1$.

Proof: It suffices to consider the case $s < 1/2$. Suppose this result is false. Let π be projection onto a vertical line. Note that Λ_s^* is a closed subset of the interior of X_s. For this reason, $\pi(\Lambda_s^*)$ either contains a segment or else is nowhere dense. So, by assumption $\pi(\Lambda_s^*)$ is nowhere dense.

Let L denote the left edge of X_s. Let L' denote the right edge of the central tile of Δ_s. Note that both L and L' have end points in the top and bottom of X_s. Hence $\pi(L) = \pi(L')$. Our argument in Lemma 16.5 shows that

$$\pi(\Lambda_s^0) = \Pi(X_s) = \pi(L).$$

Since $\pi(\Lambda_s^*)$ is nowhere dense, we must have

$$\pi(\Lambda_s^0 \cap L) = \pi(L).$$

Since $\Lambda_s^0 \cap L$ is closed, we must have $L \subset \Lambda_s^0$. But this is absurd. By the Main Theorem, at least one tile of Δ_s^0 has an edge contained in L. □

Remarks:
(i) Our argument for Lemma 16.5 breaks down in the rational case, because of the existence of triangular tiles.
(ii) We will see in Part 4 of the monograph that Λ_s is a Cantor set for almost all $s \in (0,1)$. Nonetheless, these Cantor sets always project onto line segments in certain directions.

16.4. Finite Unions of Lines

Now we prove that Λ_s cannot be contained in a finite union of lines, when s is irrational. This is Statement 3 of Theorem 1.5.

Suppose that Λ_s is contained in a finite union of lines. By Corollary 16.7, one of these lines L must be such that $L \cap \Lambda_s$ contains a line segment S that is disjoint from all the other lines in our finite list. By the Covering Lemma, we can find a finite union of patches which cover S and which have diameter much smaller than the distance from S to any other line. This implies that there exists another irrational parameter t, and a symmetric piece K_t, such that $K_t \cap \Lambda_t$ is contained in a single line segment Σ. Inspecting Lemmas 14.5 and 14.6, we see that, without loss of generality, we can assume that $K_t \neq P_t$ if $t < 1/2$. The point is that P_t contains patches of other types in these cases.

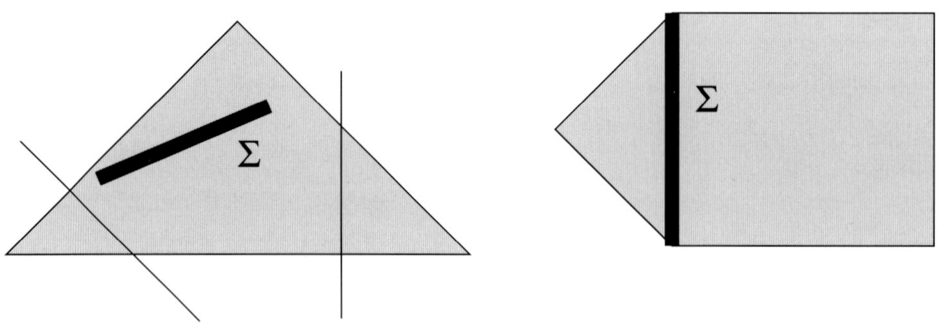

Figure 16.1: Good lines which avoid Σ.

Say that a line is good if it travels in the direction of some 8th root of unity. Suppose first that K_t is an isosceles triangle. The same argument as in Lemma 16.5 implies every good line connecting a short edge of K_t to the hypotenuse much intersect Σ. But this is only possible is Σ is the hypotenuse and moreover every point of Σ has a neighborhood in K_t which contains infinitely many tiles. This contradicts Lemma 15.2.

The other possibility is that K_t is a pentagon. Since $K_t \neq P_t$, we must have $K_t = A_t$. In this case, K_t is similar to the kind of pentagon shown on the right hand side of Figure 16.1. The only way out of the contradiction above is that Σ is the line segment shown and moreover $\Sigma \subset \Lambda$. However, we can apply the reflection R_D in the line of diagonal symmetry for P_t. The line Σ crosses D_t and hence $R_D(\Sigma \cap P_t)$ is contained in Λ_t and intersects the interior of A_t. This is a contradiction.

16.5. Existence of Aperiodic Points

Here we prove that Λ'_s is dense in Λ_s when s is irrational. This is Statement 4 of Theorem 1.5.

LEMMA 16.8. *Let s be any irrational parameter and Let $p \in \Lambda_s$ be some point and let D be any disk centered at p. Then $D \cap \Lambda_s$ is not contained in a finite union of lines.*

Proof: If this lemma is false, then we can further shrink D so that $\Lambda_s \cap D$ is contained in a single line. (This uses the fact that Λ_s has no isolated points, so we can "focus our attention" away from the intersections of the lines.) But then we can find a small patch (K_u, ψ, u) such that $K' = \psi(K_u)$ is contained in D. But then $K' \cap \Lambda_s$ is contained in a line segment. This gives us the same contradiction as in the previous section. □

LEMMA 16.9. *Let s be any irrational parameter. Given any $p \in \Lambda_s$ and any $N > 0$ and any $\epsilon > 0$ there is a point $q \in \Lambda_s$ such that $\|p - q\| < \epsilon$ and the first N iterates (forward and backward) of f_s are defined on q.*

Proof: Let D denote the disk of radius ϵ about p. The set of points of D, for which f_s^N is not defined, is contained in a finite union of lines. Now we apply the previous result. □

Now we prove that Λ'_s is dense in Λ. Choose $p \in \Lambda_s$ and let $\epsilon > 0$ be given. We set $\epsilon_1 = \epsilon/4$ and let $q_1 \in \Lambda_s$ be some point such that $\|p - q_1\| < \epsilon_1$ and $f_s^{\pm 1}$ is defined on q_1. Assume that $q_k \in \Lambda_s$ has been chosen in such a way that $f_s^{\pm k}$ is defined on p_k. There is some $\epsilon'_k > 0$ so that $f_s^{\pm k}$ is defined on all points within ϵ'_k of q_k. Let U_k denote the ϵ'_k-neighborhood of q_k.

We choose

$$\epsilon_{k+1} = \min(\epsilon_1, ..., \epsilon_k, \epsilon'_k)/4. \tag{16.2}$$

By Lemma 16.9 we can choose $q_{k+1} \in \Lambda_s$ so that $\|q_{k+1} - q_k\| < \epsilon_{k+1}$ and $f_s^{\pm(k+1)}$ is defined on q_{k+1}. By construction $\{q_k\}$ is a Cauchy sequence. The limit $q = \lim q_k$ is within $\epsilon/2$ of p and lies in U_k for all k. Hence f_s^k is defined on q for all $k \in \boldsymbol{Z}$. Note that $q \in \Lambda_s$ because Λ_s is closed. Hence $q \in \Lambda'_s$. Since ϵ is arbitrary, we see that Λ'_s is dense in Λ_s.

16.6. Hyperbolic Symmetry

In this section, we prove Theorem 1.10, which says that (X_s, f_s) and (X_t, f_t) are locally equivalent when s and t are in the same orbit of the $(2, 4, \infty)$ triangle group Γ.

Let $i = \sqrt{-1}$, as usual. Let Γ' denote the group of maps of $\boldsymbol{C} \cup \infty$ generated by the following maps.

$$z \to \overline{z}, \qquad z \to -z, \qquad z \to z - 1, \qquad z \to 1/(2z). \tag{16.3}$$

Let $\Gamma \subset \Gamma'$ be the index 2 subgroup of Γ' which preserves $\boldsymbol{H}^2$.

LEMMA 16.10. *If s and t lie in the same Γ' orbit, then (X_s, f_s) and (X_t, f_t) are locally equivalent.*

Proof: Local equivalence is an equivalence relation, so we just have to check this result on the generators of Γ'. Complex conjugation fixes $\boldsymbol{R}$ pointwise. So, for this generator there is nothing to prove. If $t = -s$ then the two systems are identical, by definition. The case $t = 1/2s$ is the Inversion Lemma.

The one nontrivial case is when $t = s - 1$. By symmetry, it suffices to consider the case when $s > 0$. There are several cases to consider. When $s > 2$, the result follows from the Insertion Lemma. Suppose that $s \in (1, 2)$. Let $s' = 1/(2s)$. Then $t = R(s')$. Combining the Covering Lemma and the Inversion Lemma, we see that X_s^0 is covered by finitely many patches of the form (K_t, ψ, t). The local equivalence follows immediately from this fact. To make it work, we need to throw out the finitely many lines containing the boundaries of the patch images.

Suppose $s \in (1/2, 1)$ and $t' = s - 1$. Then $t' < 0$ and we can switch to $t = -t' = 1 - s$. Now we apply the Covering Lemma, just as in the previous case. Finally, suppose $s \in (0, 1/2)$. Again, we consider $t = 1 - s \in (1/2, 1)$. Switching the roles of s and t we reduce to the previous case. □

It just remains to recognize Γ. The elements

$$z \to -\overline{z}, \qquad z \to \overline{1 - z}, \qquad z \to 1/\overline{2z} \tag{16.4}$$

all belong to Γ, and are the hyperbolic reflections in the 3 sides of the $(2, 4, \infty)$ triangle with vertices $i\sqrt{2}$, $(1 + i)/2$ and ∞. We omit the proof that these elements generate Γ, since what we have done already gives a proof of Theorem 1.10.

CHAPTER 17

Hausdorff Convergence

17.1. Results about Patches

In this chapter we prove Theorem 1.9, which deals with the Hausdorff continuity of the set Ξ_s as a function of s. Again, $\Xi_s = \Lambda_s$ when s is irrational, and Ξ_s is the union of the unstable orbits when s is rational. We also prove a related result, which will help establish Theorem 1.6.

LEMMA 17.1. *Suppose K_s is a symmetric piece. If s is irrational, then Λ_s contains two vertices of K_s. If s is rational, then K_s contains two triangles τ_1 and τ_2 which abut non-adjacent edges of K_s.*

Proof: This has the same proof as Lemma 16.2. □

Our discussion now refers to the Covering Lemma of §14. There are two natural sequences which arise in connection with any irrational parameter s:

- $\lambda_1, \lambda_2, \ldots$ is the set of scale factors of the maps which arise in Theorem 12.1.
- The proof of the Covering Lemma yields constants $\epsilon_1, \epsilon_2, \ldots$ such that all the periodic tiles in the patch covering corresponding to λ_k have diameter at least ϵ_k.

Given any parameter s and any set $D \subset X_s$ (typically a disk) we define the *patch spectrum* $\Pi(D, s) \subset \boldsymbol{R}$ to be the set of ϵ such that there is a symmetric ϵ-patch (K, ψ, u) such that $\psi(K) \subset D$.

LEMMA 17.2. *Let $s \in (0, 1)$ be arbitrary. Suppose that D_1 and D_2 are concentric disks, having radius ρ and 2ρ respectively. Suppose that D_1 contains a tile of diameter less than ϵ_k. Then $\lambda_k \in \Pi(D_2, s)$ provided that $2\lambda_k < \rho$.*

Proof: It follows right from the definition of the sequence $\{\epsilon_k\}$ that the λ_k covering must have a patch whose image intersects D_1. But the radius condition guarantees that the entire patch must be contained in D_2. □

17.2. Convergence

In this section we prove Theorem 1.9. Suppose that $\{s_n\}$ is a sequence of parameters converging to an irrational parameter s. We first note a continuity property. The sequences $\{\lambda_{s_n,k}\}$ and $\{\epsilon_{s_n,k}\}$ (which are either infinite sequences or finite sequences of growing length) converge pointwise to the sequences $\{\lambda_{s,k}\}$ and $\{\epsilon_{s,k}\}$ as $n \to \infty$. This is a consequence of the convergence of all the relevant sets involved in the Covering Lemma. We call this convergence *stabilization*.

Let $\Lambda = \Xi_s$. Let $L\Lambda$ be the set of limits of sequences of the form $\{p_n\}$ where $p_n \in \Xi_{s_n}$. It follows from compactness that $L\Lambda$ is the Hausdorff limit of $\{\Xi_{s_n}\}$. So, to finish the proof of Theorem 1.9, we just need to prove that $L\Lambda = \Lambda$.

LEMMA 17.3. *$L\Lambda$ is disjoint from the interiors of the tiles of Δ.*

Proof: Let $p = \lim p_n$ with $p_n \in \Lambda_n$. Suppose that p lies in the interior of a tile P of Δ. Then there is some tile P_n of Δ_n such that $P_n \to P$ as $n \to \infty$. There is some disk centered at p which is contained in P_n for all n large. But then $p_n \in P_n$ for n large, and this is a contradiction. □

LEMMA 17.4. *$L\Lambda$ is disjoint from the interiors of edges which are common to two tiles of Δ.*

Proof: Let $p = \lim p_n$ with $p_n \in \Lambda_n$. Suppose p lies in the interior of the edges of two tiles P and P^* of Δ. Let P_n and P_n^* be the two tiles of Δ_n converging to P and P^* respectively. By Lemma 8.11, the union $P_n \cup P_n^*$ eventually contains a uniformly large neighborhood of p. But then we get the same contradiction as in the previous case. □

LEMMA 17.5. *$L\Lambda$ has no isolated points.*

Proof: Let $p \in L\Lambda$. Let D_2 be a closed disk centered at p. Let D_1 be the concentric disk having half the radius. Let δ_n be the infimal diameter of a tile in $\Delta_n \cap D_1$. When n is sufficiently large, D_1 contains a tile of Ξ_{s_n}. Hence $\delta_n \to 0$ as $n \to \infty$.

Consider the patch spectra $\Pi(D_2, s_n)$. By the stabilization property and Lemma 17.2, there is some $\delta > 0$ so that

$$\Pi(D_2, s_n) \cap [\delta, 1] \neq \emptyset$$

for all n sufficiently large.

By 17.1, there are two elements (either points or unstable tiles) of Ξ_{s_n}, uniformly separated from each other and contained in D_2. Hence $L\Lambda \cap D_2$ contains at least 2 points. But the size of D_2 is arbitrary. Hence, in the irrational case, there cannot be any isolated points. □

LEMMA 17.6. *$L\Lambda = \Lambda$.*

Proof: Suppose $p \in L\Lambda - \Lambda$. From the results we have proved above, p must be the vertex of a tile of Δ, and moreover, some neighborhood of p intersects only finitely many tiles. But then p is an isolated point of $L\Lambda$, and this contradicts the previous result. Hence $L\Lambda \subset \Lambda$.

To show that $\Lambda \subset L\Lambda$ we choose a point $p \in \Lambda$ and run exactly the same argument as in the proof of Lemma 17.5. What we get is that every disk D_2 centered at p contains an element of Ξ_{s_n} for all sufficiently large n. □

COROLLARY 17.7. Let $\{s_n\}$ be an irrational sequence converging to an irrational parameter s. The limit sets $\Lambda_{s_n} \to \Lambda_s$ in the Hausdorff topology.

Proof: Let $\Lambda = \Lambda_s$ and $\Lambda_n = \Lambda_{s_n}$. For any $\epsilon > 0$ there is some n such that Λ_n is contained in the ϵ tubular neighborhood of Λ once n is large. Otherwise, we could find a sequence $\{p_n\}$ which avoided the ϵ tubular neighborhood. But then we could extract a convergent subsequence and produce a point of $L\Lambda - \Lambda$.

Conversely, suppose that Λ is not contained in the ϵ tubular neighborhood of Λ_n no matter how large n is. Then some sequence $\{p_n\}$ of Λ is such that p_n avoids the ϵ-tubular neighborhood of Λ_n. Any subsequential limit of this sequence belongs to $\Lambda - L\Lambda$. □

17.3. Covering

In this section we prove a result which is a precursor to Theorem 1.6. We call σ a *nice rational* if $\sigma \neq 1/2n$ for $n = 1, 2, 3, \ldots$ To avoid annoying special cases, we will only work with nice rationals.

We described in §11.5 what it means to say that $\sigma \to s$ when $\sigma, s < 1/2$. Essentially it means that the even expansion of s is a continuation of the even expansion of σ, except perhaps that the last two digits of σ need to be altered. Here we extend this definition to $\sigma, s \in (1/2, 1)$. We write $\sigma \to s$ in this case if and only if $R(\sigma) \to R(s)$ in the sense of §11.5. Here $R(s) = 1 - s$ and $R(\sigma) = 1 - \sigma$. With this definition, the Diophantine estimate in Lemma 11.5 applies immediately to all $\sigma, s \in (0, 1)$ such that $\sigma \to s$.

Given a polygon P, let CP denote the polygon obtained from P by dilating P by a factor of C about the center of mass of P. Given any K, standing for one of the symmetric pieces $\{A, B, P, Q\}$, we define

$$\Xi(\sigma, K, C) = \bigcup_{\tau \in \Xi_\sigma \cap K_\sigma} C\tau. \tag{17.1}$$

We are taking the union of inflated versions of the triangular tiles in K_σ. We define $\Xi(\sigma, C)$ in a similar way, except that we inflate all the tiles of Ξ_σ.

THEOREM 17.8. *Suppose σ is a nice rational and $\sigma \to s$. Then there is some constant C such that $\Xi(s, K, 0) \subset \Xi(\sigma, K, C)$ for each symmetric piece K. The constant C does not depend on the parameters or on the symmetric piece.*

COROLLARY 17.9. Suppose σ is a nice rational and $\sigma \to s$. Then there is some constant C such that $\Xi_s \subset \Xi(\sigma, C)$ The constant C does not depend on the parameters or on the symmetric piece.

We prove Theorem 17.8 though a series of smaller lemmas.

LEMMA 17.10. *Theorem* 17.8 *holds when $\sigma = (k-1)/k$ for $k = 2, 4, 6, \ldots$.*

Proof: Here σ has even expansion $(1, k)$. For any particular choice of k, the result follows just because there is a uniform lower bound to the size of the tiles of Δ_σ. We just have to understand what happens when $k \to \infty$. The situation here is the same one depicted in the proof of the Area Lemma. Figure 17.1 illustrates the situation for the symmetric piece P. The other cases are similar.

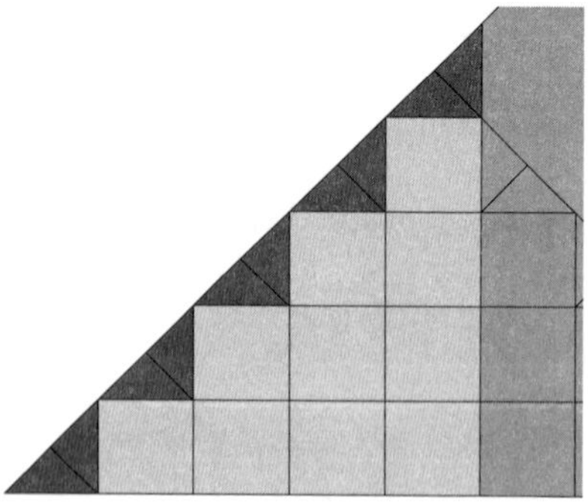

Figure 17.1: P_σ (white/dark) and $\Xi(\sigma, P, 0)$ (dark)

Figure 17.2 shows $\Xi(s, Q, 0)$ for $s = 29/32$ and $s = 17/19$. The even expansion in the first case is $(1, 10, 2, 2)$. In the second case, the even expansion is $(1, 9, 1, 4)$. The color scheme is the same as in Figure 17.1.

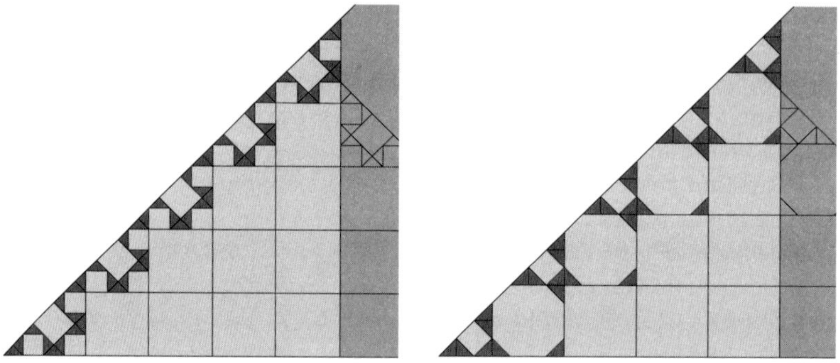

Figure 17.2: Δ_s^0 for $s = 29/32$ and $s = 17/19$.

We will make a general argument but we will refer to the figures above for visual guidance. We will give our proof for the symmetric piece P_s, which is the triangular region colored light and dark grey in the figures. (The region colored medium grey is not part of the piece.) The proofs for the other cases are similar. We note the following:

- P_s is within C/k^2 of P_σ. This follows from Lemma 11.5 and from the definitions of these sets.
- $\Xi(s, P, 0)$ is contained in the $4/k$ tubular neighborhood of the diagonal edge of P_s. This follows from Lemma 15.8 and Lemma 11.5.
- The diagonal edge of P_σ is covered by edges of tiles of Ξ_σ, and each such tile has short side length $1/k$.

From these properties, the set $\Xi(\sigma, P, 100)$ covers the $4/k$ tubular neighborhood of the diagonal edge of P_s. Hence $\Xi(s, P, 0) \subset \Xi(\sigma, P, 100)$. $\square$

LEMMA 17.11. *Theorem 17.8 holds when* $\sigma = (k-1)/k$ *or* $\sigma = (k-1)/2k$ *for* $k = 2, 3, 4, \ldots$.

Proof: The cases not covered by the preceding result are covered by essentially the same argument as in the preceding result. The proof boils down to Lemmas

11.5 and 15.8. □

The rest of our proof is an induction argument. Let $\lambda(\sigma)$ denote the length of the even expansion of σ. The cases $\lambda(\sigma) = 1$ correspond to $\sigma = 1/2k$, and we have excluded these from consideration. Lemma 17.11 takes care of all the cases when $\lambda(\sigma) = 2$. We say that a constant C_m is *m-good* if Theorem 17.8 holds for all σ, with the constant C_m, provided that $\lambda(\sigma) \le m$. We have already shown the existence of a 2-good constant C_2.

LEMMA 17.12. *If C_m is an m-good constant, then*

$$C_{m+1} = C_m + \frac{\Omega}{\sqrt{2}^m}$$

is an $(m+1)$-good constant. Here Ω is independent of m.

Proof: For ease of notation we give the proof for the symmetric piece A_s. Let $t = R(t)$ and $\tau = R(\sigma)$. Here R is as in the Main Theorem. The map R just cuts off the first digit of the even expansions, so we have $\tau \to t$. Let K_s be some symmetric piece. We want to show $\Xi(s, A, 0) \subset \Xi(\sigma, A, C_{m+1})$.

Since $\lambda(\tau) = m$, we have

$$\Xi(t, K, 0) \subset \Xi(\tau, K, C_m), \qquad \forall\, K \in \{A, B, P, Q\}. \tag{17.2}$$

By the Covering Lemma, we know that

$$\Xi(s, A, 0) = \bigcup_{k=1}^{n} F_{k,s}(\Xi(t, K_k, 0)). \tag{17.3}$$

The constant n depends on the equations used when we proved the Covering Lemma. The map $F_{k,s}$ is a patch map and $K_{k,s}$ is a symmetric piece whose type depends on the index k. Combining the last two equations, we get

$$\Xi(s, A, 0) \subset \bigcup_{k=1}^{n} F_{k,s}\Big(\Xi(\tau, K_k, C_m)\Big). \tag{17.4}$$

Since $\lambda(\sigma) \ge 3$, the first two terms in the even expansion of σ are the same as the first two terms in the even expansion of s. Hence σ and s fall in the same case of our proof of the Covering Lemma, and they have the same layering constant. For this reason, the patch covering of A_s has the same combinatorics as the patch covering of A_σ. Hence, for any constant C,

$$\Xi(\sigma, K_\sigma, C) = \bigcup_{k=1}^{n} F_{k,\sigma}\Big(\Xi(\tau, K_k, C)\Big). \tag{17.5}$$

Here the maps $F_{k,s}$ and $F_{k,\sigma}$ have the same combinatorial structure. They are compositions of the map ϕ from the Main Theorem, the map T^j from the Filling Lemma, and one or more of the fundamental symmetries. For example,

$$F_{1,s} = R_D \circ T_s^2 \circ \phi_s \qquad \Longleftrightarrow \qquad F_{1,\sigma} = R_D \circ T_\sigma^2 \circ \phi_\sigma.$$

Let D_τ denote the denominator of τ. Given the identical combinatorial structure, it follows from Lemma 11.5 and elementary geometry that

$$\|F_{k,s}(p) - F_{k,\sigma}(p)\| \le \frac{\Omega'}{(\text{denominator}(\tau))^2}, \tag{17.6}$$

for all points p contained in (say) the disk of radius 100 centered at the origin. Here Ω' is some constant that does not depend on any of the parameters. The point is that the relevant polygons in the domain are $O(D_\sigma^{-2})$ apart in the Hausdorff metric, and the corresponding polygons in the range are $O(D_\tau^{-2})$ apart, and the maps in question are 1-lipschitz.

Finally, the short side length of each tile θ of U_τ is exactly $1/D_\tau$. Hence

$$F_{k,s}\Big(\Xi(\tau, K_k, C_m)\Big) \subset F_{k,\sigma}\Big(\Xi(\tau, K_k, C_m + 100\Omega'/D_\tau)\Big), \tag{17.7}$$

We have added the factor of 100 to avoid tedious Euclidean geometry. The estimate $D_\tau \geq (\sqrt{2})^m$. This gives

$$F_{k,s}\Big(\Xi(\tau, K_k, C_m)\Big) \subset F_{k,\sigma}\Big(\Xi(\tau, K_k, C_{m+1})\Big). \tag{17.8}$$

Equations 17.4, 17.5, and 17.8 combine to finish the proof, provided that we set $\Omega = 100\Omega'$. $\square$

The constant

$$C = C_2 + \sum_{m=2}^{\infty} \frac{\Omega}{\sqrt{2}^m} \tag{17.9}$$

works in Theorem 17.8. This completes the proof of Theorem 17.8.

CHAPTER 18

Recurrence Relations

Now we revisit the proof of the Covering Lemma to deduce some recurrence relations for the period $N(s)$ of the unstable orbits at parameter s. We found these rules experimentally and below we will give proofs.

We set $N(0) = 0$. By inspection, $N(1/2) = 1$. By the Insertion Lemma, we have $N(1/2k) = 1$ for $k = 1, 2, 3, \ldots$. Let $t = R(s)$ and $u = R(t)$. The notation $(a, b) \to (c, d)$ has the following meaning. If $(a, b, \ldots)$ is the even expansion of s then

$$N(s) = cN(t) + dN(u) \tag{18.1}$$

We will establish the following rules:

(1) $(a, b) \to (b, 1)$ for a and b even.
(2) $(a, 1) \to (1, 1)$ for $a > 1$ odd.
(3) $(1, b) \to (b - 3, 3)$ for $b > 1$ odd.
(4) $(1, b) \to (b - 1, 1)$ for b even
(5) $(a, b) \to (b - 2, 3)$ for a even and $b > 1$ odd.

These rules allow one to compute $N(s)$ recursively for any rational s. Here is an example. $7/18$ has even expansion $(2, 3, 1, 4)$. Let $[2, 3, 1, 4]$ denote $N(7/18)$, etc. We have

- $[2, 3, 1, 4] = [3, 1, 4] + 3[1, 4]$ by Rule 5.
- $[3, 1, 4] = [1, 4] + [4]$ by Rule 2.
- $[1, 4] = 3[4]$ by Rule 4.

Hence $N(7/18) = [2, 3, 1, 4] = 13[4] = 13(N(1/4)) = 13$.

The proofs of all the rules boil down to inspecting the decompositions we used when proving the Covering Lemma – and occasionally improvising some new related ones.

Let $N(A_s)$ denote the number of triangular tiles in $A_s \cap \Delta_s$, etc. We have

$$2N(s) = N(A_s) + N(B_s) = N(P_s) + N(Q_s) \tag{18.2}$$

Equation 18.2 holds for any parameter.

LEMMA 18.1. *If $s, t < 1/2$ and $N(A_t) = N(B_t)$, then $N(A_s) = N(B_s)$.*

Proof: Figures 14.3 and 14.8 illustrate the partition

$$A_s = R_D(B_s - Q_s) \cup R_V \phi_s(A_t) \cup \text{squares} \tag{18.3}$$

It follows from this partition that

$$N(A_s) - N(B_s) = N(A_t) - N(Q_s). \tag{18.4}$$

But $Q_s = \phi_s(B_t)$ by Lemma 14.5, Equation 4. Hence

$$N(A_s) - N(B_s) = N(A_t) - N(B_t). \tag{18.5}$$

The lemma follows immediately from this last equation. □

LEMMA 18.2. *If $s > 1/2$ and $N(A_t) = N(B_t)$, then*
$$N(P_s) - N(Q_s) = 2N(A_s).$$

Proof: Lemmas 14.3 and 14.4 imply that

$$N(A_s) = N(B_t). \tag{18.6}$$

As illustrated in Figures 14.1, 14.2, 14.6, and 14.7, we have the partition

$$P_s = R_V(Q_s) \cup R_D(\Psi_s^0). \tag{18.7}$$

Here R_V and R_D are reflections in two of the fundamental lines of symmetry. Equation 18.7 gives us

$$N(P_s) - N(Q_s) = N(\Psi_s^0) = N(A_t) + N(B_t) = 2N(B_t) = 2N(A_s). \tag{18.8}$$

The last equality comes Equation 18.6. □

Our proof of the recurrence relations follows from the decompositions given in the proof of the Covering Lemma and the following result.

LEMMA 18.3. *Let $s \in (0,1)$ be a rational parameter.*
- *If $s \le 1/2$, then $N(A_s) = N(B_s)$.*
- *If $s > 1/2$ then $N(P_s) - N(Q_s) = 2N(A_s)$.*

Proof: By Lemma 18.2, Statement 1 of Lemma 18.3 implies Statement 2 of Lemma 18.3. So, we will prove Statement 1.

We check that $N(A_s) = N(B_s) = 1$ when $s = 1/2$. The same result holds for $s = 1/2n$ by the Insertion Lemma. Let $t = R(s)$. Our proof goes by induction in the integer k such that $R^k(s) = 0$. If Lemma 18.3 fails, we can choose a minimal counterexample s. Let $t = R(s)$. It follows from Lemma 18.1 and minimality that $t > 1/2$.

Examining the equations given in the proof of Lemma 14.6, we see that

$$N(A_s) = 2N(A_t) + N(Q_t), \qquad N(B_s) = N(P_t). \tag{18.9}$$

Therefore

$$N(B_s) - N(A_s) = N(P_t) - N(Q_t) - 2N(A_t). \tag{18.10}$$

Applying Lemma 18.2 to the parameter $t > 1/2$ and using the minimality of s, we see that the right hand side of Equation 18.10 vanishes. □

Proof of Rule 1: In this case a and b are even, and we want to prove that $N(s) = bN(t) + N(u)$. By the Filling Lemma from §13, and symmetry, the number of triangular tiles in Ψ_s^k is $2N(t)$ for $k = 0, ..., (\mho - 1)$. Here $\mho = \mho(s) = b/2$ is the layering constant. By Lemma 18.3, we have

$$N(s) = N(A_s) = N(B_s). \tag{18.11}$$

Statement 2 of Lemma 14.5 gives us the identity

$$N(B_s) = 2\mho N(t) + N(Q_t) = bN(t) + N(Q_t). \tag{18.12}$$

Since a and b are even, t also satisfies the hypotheses of Lemma 14.5. Hence, by Equation 4 of Lemma 14.5,

$$N(Q_t) = N(B_u). \tag{18.13}$$

These equations together imply Rule 1. □

Proof of Rule 2: Suppose $a > 1$ is odd and $b = 1$. We prove that $N(s) = N(t) + N(u)$. Here $s, t < 1/2$ and $\mho(s) = 1$.

Equations 1 and 2 of Lemma 14.6 combine with Equation 18.2 to give

$$2N(s) = N(A_s)+N(B_s) = (2N(A_t)+N(Q_t))+N(P_t) = 2N(t)+2N(A_t). \tag{18.14}$$

Equation 1 of either Lemma 14.3 or 14.4, as applied to the parameter t, gives us

$$2N(A_t) = 2N(B_u) =^* N(A_u) + N(B_u) = 2N(u). \tag{18.15}$$

The starred equality comes from Lemma 18.3. Rule 2 follows from the Equations 18.14 and 18.15. □

Proof of Rule 3: In this case, $a = 1$ and $b > 1$ is odd. We show $N(s) = (b-3)N(t) + 3N(u)$. We have $s > 1/2$ and $\mho(s) = (b-1)/2$. In particular $2(\mho(s) - 1) = b - 3$.

Since $s > 1/2$, Statement 2 of Lemma 18.3 implies

$$2N(s) = N(P_s) + N(Q_s) = 2N(A_s) + 2N(Q_s).$$

In short,

$$N(s) = N(A_s) + N(Q_s). \tag{18.16}$$

Statement 4 of either Lemma 14.3 or 14.4 gives

$$N(Q_s) = (b-3)N(t) + N(Q_t), \qquad N(A_s) = N(B_t). \tag{18.17}$$

Note that $t < 1/2$ and $\mho(t) = 1$. Equations 2 and 4 of Lemma 14.5, applied to the parameter t, give

$$N(B_t) = N(P_u), \qquad N(Q_t) = N(B_u). \tag{18.18}$$

Combining these equations, we get

$$N(s) = (b-3)N(t) + N(P_u) + N(B_u). \tag{18.19}$$

Combining this information with Equation 18.2, we get

$$N(B_u) + N(P_u) = (2N(u) - N(A_u)) + (2N(u) - N(Q_u)) = 3N(u). \tag{18.20}$$

Rule 3 follows from Equations 18.19 and 18.20. □

Proof of Rule 4: In this case, $a = 1$ and $b > 1$ is even. We show $N(s) = (b-1)N(t) + N(u)$. We have $\mho(s) = b/2$ and $s > 1/2$. Since b is even, $t < 1/2$ and $u < 1/2$.

Equation 4 of Lemma 14.3 or Lemma 14.4 combines with Equation 18.16 to give

$$N(s) = (b-2)N(t) + N(B_t) + N(Q_t). \tag{18.21}$$

Since $t < 1/2$, Lemma 18.3 gives

$$N(B_t) = N(t). \tag{18.22}$$

Hence

$$N(s) = (b-1)N(t) + N(Q_t). \tag{18.23}$$

Statement 4 of Lemma 14.5 or 14.6, applied to the parameter t, gives

$$N(Q_t) = N(B_u) = N(u). \tag{18.24}$$

Rule 4 follows from the last two equations. □

Proof of Rule 5: We have a even and $b > 1$ odd. We show $N(s) = (b-2)N(t) + 3N(u)$. Here $s < 1/2$ and $t < 1/2$ and $u > 1/2$. Here $\mho(s) = (b+1)/2$. We get the same decomposition as what we had for Rule 2, except that the sum only goes up to $\mho - 2$. This gives us the equation

$$N(s) = (b-1)N(t) + N(Q_t). \tag{18.25}$$

$\mho(t) = 1$, so Equation 4 of Lemma 14.6 gives

$$N(Q_t) = N(B_u). \tag{18.26}$$

Hence

$$N(s) = (b-1)N(t) + N(B_u) = (b-2)N(t) + N(A_t) + N(B_u). \tag{18.27}$$

The second equality comes from Lemma 18.3. Equation 1 of Lemma 14.6 gives

$$N(A_t) = 2N(A_u) + N(Q_u). \tag{18.28}$$

Putting everything together, we get

$$N(s) = (b-2)N(t) + 2N(A_u) + N(Q_u) + N(B_u) =$$

$$(b-2)N(t) + 2N(u) + N(A_u) + N(Q_u) = (b-2)N(t) + 3N(u). \tag{18.29}$$

The last equality comes from Equation 18.16 applied to $u > 1/2$. □

CHAPTER 19

Hausdorff Dimension Bounds

19.1. The Upper Bound Formula

Our goal in this chapter is to prove Theorem 1.6. We adopt the notation $\sigma \to s$ in the sense of Theorem 17.8. This means that σ approximates s in the sense of the even expansions. We first formulate and prove a more precise version of the first statement of Theorem 1.6.

LEMMA 19.1. *Let $s \in (0,1)$ be irrational. Suppose that $\{\sigma_n\}$ is a sequence of rationals such that $\sigma_n \to s$ in the sense of Theorem 17.8. Then*

$$\dim(\Lambda_s) \leq \limsup \frac{\log N(\sigma_n)}{\log D(\sigma_n)}. \tag{19.1}$$

Here $N(\sigma_n)$ is the common period of the unstable orbits associated to the parameter σ_n and $D(\sigma_n)$ is the denominator of σ_n.

Proof: By Theorem 17.8, we have

$$\Lambda_s \subset \Xi(\sigma_n, C') \qquad \forall n \tag{19.2}$$

for some constant C'. Set $C = 2C'$.

Let D be the quantity on the right hand side of Equation 19.1. We fix n for the moment. Let $m = q_n$ and $N_n = N(\sigma_n)$. Here N_n is the number of polygons in the cover associated to σ_n. The sets in $\Xi(\sigma_n, C)$ are triangles having diameter at most C/m. For any given $\epsilon > 0$ we can choose n large enough so that

$$\log(N_m) < (D+\epsilon)\log(m). \tag{19.3}$$

But then $N_m < m^{D+\epsilon}$. By Lemma 2.9, we have $\dim(\Lambda_s) \leq D + \epsilon$. Since ϵ is arbitrary, we get $\dim(\Lambda_s) \leq D$. □

The rest of the proof of Theorem 1.6 has nothing specifically to do with the octagonal PETs. We just make some estimates on the right hand side of Equation 19.1.

19.2. A Formula in the Oddly Even Case

In this section we derive the following result from Rule 1 listed in §18.

THEOREM 19.2. *Suppose $s \in \mathbf{Q}$ has has tweaked C.F.E. $(0, a_1, , ...a_n)$, as in §11.3, where a_k is even for all odd k and there are at least 2 nonzero terms. Then $N(s)$ is the denominator of the fraction with C.F.E. $(0, 2a_2, a_3, 2a_4, a_5, 2a_6, a_7...)$ with $n-1$ nonzero terms.*

Proof: We claim that Theorem 19.2 holds when $s = 1/m$ for $m = 2, 3, 4, \ldots$ We check that $N(1/m) = 1$ and $N(1/m) = 2$ when $m = 2$ and $m = 3$ respectively. It follows from the Insertion Lemma that $N(1/m) = 1$ if m is even and $N(1/m) = 2$ if m is odd. When m is even, the even expansion has only 1 term, and our result does not apply. When m s odd, the tweaked C.F.E. of $1/m$ is $(0, m-1, 1)$. The C.F.E. predicted by our result is $(0, 2)$, and this is the C.F.E. for $1/2$. So, the result works in this case.

Now suppose that $s \neq 1/k$ is an oddly even rational. Let $(0, a_1, a_2, \ldots)$ be the continued fraction expansion of s. By Lemma 11.4, the even expansion of s is $(a_1, 2a_2, a_3, 2a_4, \ldots)$.

Now, let $s_m = R^m(s)$ for $m = 0, 1, 2, \ldots$ Applying Rule 1 from §18 repeatedly, we have

- $N(s_0) = 2a_2 N(s_1) + N(s_2)$,
- $N(s_1) = a_3 N(s_2) + N(s_3)$,
- $N(s_2) = 2a_4 N(s_3) + N(s_4)$,

and so on. As we discussed in §2.7, these recurrence relations imply that $N(s_0)$ is the denominator of the fraction whose continued fraction expansion is $(0, 2a_2, a_3, 2a_4, \ldots)$. $\square$

19.3. One Dimensional Examples

Now we prove that $\dim(\Lambda_s) = 1$ when $\lim_{n\to\infty} R^n(s) = 1$. Given the invariance of the dimension under the action of the renormalization map R, it suffices to consider the case when s is oddly even. Suppose that s has C.F.E. $(0, a_1, a_2, \ldots)$. Let σ be the approximating rational with continued fraction $(0, a_1, \ldots, a_n)$. We take n odd so that $\sigma \to s$ in the sense of Theorem 17.9. We have the easy bound

$$D(\sigma) > \prod_{i=1}^{n} a_i. \tag{19.4}$$

$N(\sigma)$ is the denominator of the fraction with C.F.E.

$$(0, 2a_2, a_3, 2a_4, \ldots, 2a_{n-1}, a_n). \tag{19.5}$$

We have the easy bound

$$N(\sigma) < \prod_{i=1}^{n} 4a_i. \tag{19.6}$$

But then

$$\frac{\log N(\sigma)}{\log D(\sigma)} < \frac{n \log 4 + \sum_{i=1}^{n} \log a_i}{\sum_{i=1}^{n} \log a_i} = 1 + \log 4\bigg(\frac{1}{\frac{1}{n}\sum_{i=1}^{n} \log(a_i)}\bigg). \tag{19.7}$$

Since $a_n \to \infty$, the term on the right tends to 1 as $n \to \infty$.

It now follows from Equation 1.8 that $\dim(\lambda_s) \leq 1$. We already know from Theorem 1.10 that $\dim(\Lambda_s) \geq 1$, so we must have $\dim(\Lambda_s) = 1$.

19.4. A Warm-Up Case

Our remaining goal in this chapter is to prove that $\dim(\Lambda_s) \le 1 + (\log 8/\log 9)$ in general. As a warm-up, we will consider the case when s is oddly even, and we will get a slightly better bound.

Suppose that s is oddly even. Let $\{\sigma_n\}$ be as in the previous section. The continued fraction for σ_{2n+1} is

$$(0, 2a_1, a_2, 2a_3, a_4, ..., a_{2n}, 2a_{2n+1}).$$

We set $\sigma_{2n+1} = p/q$, and we want to get a bound on $D(\sigma) = q$.

We have $q = q_0 > q_1$ and

$$q_1 = 2a_2q_2 + q_3 = 2a_2(2a_3q_3 + q_4) + q_5 \ge (2a_2a_3 + 1)q_3.$$

Similarly $q_3 \ge (2a_4a_5 + 1)q_5$, and so on. Hence

$$D(\sigma_{2n+1}) > \prod_{k=1}^{n}(2a_{2k}a_{2k+1} + 1). \tag{19.8}$$

Let $r = N(sigma_{2n+1})$ be the denominator of the fraction whose continued fraction expansion is

$$(0, 2a_2, 2a_3, 2a_4, ..., 2a_{2n}, 2a_{2n+1}). \tag{19.9}$$

We introduce the numbers $r = r_1, r_2, r_3, ...$ which obey the basic recurrence relation with respect to the new continued fraction. We have

$$r_1 = (4a_1a_2 + 2a_1)r_3 + r_4 \le (4a_1a_2 + 2a_1 + 1)r_3.$$

A similar bound holds when all the indices are shifted by $2, 4, 6, ...$ Hence

$$N(\sigma_{2n+1}) < \prod_{k=1}^{n}(4a_{2k}a_{2k+1} + 2a_{2k} + 1). \tag{19.10}$$

A bit of calculus shows that

$$\sup_{a\ge 1, b\ge 1} \frac{\log(4ab + 2a + 1)}{\log(2ab + 1)} = \log(7)/\log(3). \tag{19.11}$$

The maximum is attained at $(a, b) = (1, 1)$. Combining Equation 19.1 with Equations 19.8, 19.10, and 19.11, we see that $\dim(\Lambda_s) \le \log(7)/\log(3)$ when s is oddly even.

19.5. Most of The General Bound

In this section, we prove that $\dim(\Lambda_s) \le 1 + (\log 8/\log 9)$ for many kinds of parameters. We clean up the remaining tricky cases in the next section. By the Main Theorem, it suffices to prove our bound for $s < 1/2$. Our method is like what we did in the last section, only more complicated.

Let $\sigma = p/q$ be a rational approximation of s. When we consider the even expansion of σ, we distinguish the following kinds of sub-strings:

- Type $0, 0$: $2c, 2d$.
- Type $1, 1$: $2c + 1, 1, 2d + 1, 1$.
- Type $1, 0$: $2c + 1, 1, 2d$.
- Type $0, 1$: $2c, 2d + 1, 1$.

Here c and d are both natural numbers, and will be throughout the section. We call these sub-strings *chunks*.

We want to estimate the quantity

$$\Delta(\sigma) = \frac{\log N(\sigma)}{\log D(\sigma)}. \tag{19.12}$$

from above. We restrict the length of the even expansion of σ so that it is divided into a finite union of chunks. For instance

$$3, 1, 2, 4, 6, 5, 1, 3, 1, 2, 4$$

divides into

$$(3, 1, 2)(4, 6)(5, 1, 3, 1)(2, 4).$$

We call these chunks $\chi_1, ..., \chi_k$. We define p_j/q_j as the rational number whose even expansion is $\chi_j, ..., \chi_k$. We need to take some care to align the indices propertly, due to the (generalization of the) fact that the continued fraction expansion in Equation 19.9 is missing the a_1 term. To make the relevant numbers line up exactly, we introduce a trick.

Let p'_j/q'_j denote the rational whose even expansion is

$$\text{last}(\chi_{j-1}), \chi_j, ..., \chi_k. \tag{19.13}$$

Here $\text{last}(\chi_{j-1})$ denotes the last digit of χ_{j-1}. The chunk χ_0 is not defined, so we set $\text{last}(\chi_0) = 2$. Let

$$r_j = N(p'_j/q'_j) \tag{19.14}$$

From our recurrence rules, we have

$$\Delta(\sigma) < \frac{\log r_1}{\log q_1}. \tag{19.15}$$

To recast the argument in the previous section in the language here, we proved the following lemma.

LEMMA 19.3. *Suppose that $\chi_j = (2c, 2d)$ has type $0, 0$. Then*

$$q_j \geq Q(c,d)q_{j+1}, \qquad r_j \leq R(c,d)r_{j+1},$$

where

$$Q(c,d) = 2cb + 1, \qquad R(c,d) = 4cd + 2c + 1.$$

Moreover

$$\frac{\log R(c,d)}{\log Q(c,d)} \leq \frac{\log 7}{\log 3}.$$

Repeated applications of Lemma 19.3 establish the special case of Statement 2 proved in the previous section.

Remark: What is crucial in this lemma is that the first digit of the sequence defining r_j does not enter into the recurrence relation for r_j. In the remaining cases, the first digit does enter into the recurrence relation, but only in a mild way: Only the parity matters.

Now we consider the remaining cases. Here is the type $1, 1$ case.

LEMMA 19.4. *Suppose that* $\chi_j = (2c+1, 1, 2d+1, 1)$. *Then*

$$q_j \geq Q(c,d)q_{j+1}, \qquad r_j \leq R(c,d)r_{j+1},$$

where

$$Q(c,d) = 2cd + 2d - 1, \qquad R(c,d) = 8cd + 8d - 2c - 8.$$

Moreover,

$$\frac{\log R(c,d)}{\log Q(c,d)} \leq \frac{\log 6}{\log 3}.$$

Proof: We will consider the case of χ_1 first. The relevant string here is $(2, 2c+1, 1, 2d+1, 1)$. The signed C.F.E. associated to this string is

$$(\overline{2c+2}, \overline{d+1}). \tag{19.16}$$

This leads to the relation

- $q_1 = (2c+2)\theta_1 - q_2$
- $\theta = (d+1)q_2 - \theta_2$.

Here θ_1 and θ_2 are dummy variables. Eliminating θ_1, we get

$$q_1 = Q(c,d)q_2 - (2+2c)\theta_2. \tag{19.17}$$

Here $Q(c,d)$ is as in the statement of the lemma. Since $\theta_2 \geq 0$, we have $q_1 \geq Q(c,d)q_2$, as claimed.

To compute r_1, the relevant string is $(2, 2c+1, 1, 2d+1, 1)$. We have the following recurrence relations:

- $r_1 = (2c-1)\rho_1 + 3\rho_2$ (Rule 5)
- $\rho_1 = \rho_2 + \rho_3$ (Rule 2)
- $\rho_2 = (2d-2)\rho_3 + 3\rho_4$ (Rule 3)
- $\rho_3 = r_2 + \rho_4$ (Rule 2)

Simplifying, we get

$$r_1 = (4cd + 4d - 3)r_2 + (4cd + 4d - 2c - 5)\rho_4 \leq^*$$

$$(4cd + 4d - 3)r_2 + (4cd + 4d - 2c - 5)r_2 = R(c,d)r_2. \tag{19.18}$$

For the starred inequality, we used the fact that $\rho_4 \leq r_2$, and that the coefficient of ρ_4 is positive for any choice of $c, d \geq 1$.

Were we to consider χ_j instead of χ_1, the digit $\text{last}(\chi_{j-1})$ might be odd. We would then get the same recurrence, except that Rule 3 would be used in place of Rule 5 in the first step, and we would get a smaller polynomial. Thus, the even case gives an upper bound which works for both cases.

A bit of calculus establishes the last assertion of the lemma. □

LEMMA 19.5. *Suppose that* $\chi_j = (2c, 2d+1, 1)$. *Then*

$$q_j \geq Q(c,d)q_{j+1}, \qquad r_j \leq R(c,d)r_{j+1},$$

where $Q(c,d) = 2cd + 1$ *and* $R(c,d) = 8cd + 2c + 2$. *Moreover*

$$\frac{\log R(c,d)}{\log Q(c,d)} \leq \frac{\log 12}{\log 3}, \qquad c + d \geq 4 \implies \frac{\log R(c,d)}{\log Q(c,d)} \leq \frac{\log 32}{\log 7}$$

Proof: The proof is just like the previous case, but with the following changes. The derivation of $Q(c,d)$ yields the recurrence relation

$$q_1 = (2cd + 2c + 1)q_2 - 2c\theta_2 \geq Q(c,d)q_2.$$

This time we are using the fact that $\theta_2 \leq q_2$. The derivation of $R(c,d)$ yields

$$r_1 = (4cd + 4c + 1)r_2 + (4cd - 2c + 1)\rho_3 \leq R(c,d)r_2.$$

Here again we use that $\rho_3 \leq r_2$.

An easy calculus argument establishes the stated bounds. □

LEMMA 19.6. *Suppose that $\chi_j = (2c+1, 1, 2d)$. Then*

$$q_j \geq Q(c,d)q_{j+1}, \qquad r_j \leq R(c,d)r_{j+1},$$

where $Q(c,d) = 2cd + 2d - 1$ and $R(c,d) = 4cd + 2c + 4d - 1$. Moreover

$$\frac{\log R(c,d)}{\log Q(c,d)} \leq 2, \qquad c + d \geq 3 \quad \Longrightarrow \quad \frac{\log R(c,d)}{\log Q(c,d)} \leq \frac{\log 15}{\log 5}.$$

Proof: The proof is essentially the same as in the previous case. We omit the details. □

We call a chunk *good* if it does not have the form $(2,3,1)$ or $(2,5,1)$ or $(4,3,1)$ or $(2,1,3)$. Otherwise we call the chunk *good*. Of the bounds associated to good chunks, $\log(32)/\log(7) \approx 1.78$ is the largest. The lemmas above immediately combine to prove:

COROLLARY 19.7. Suppose that the even expansion of s can be decomposed into good chunks. Then

$$\dim(\Lambda_s) \leq \frac{\log 32}{\log 7}.$$

19.6. Dealing with the Exceptions

For the exceptional cases, we consider the chunks two at a time. For reference, we gather together the polynomials discussed in the previous chapter:

- $Q_{00}(c,d) = 2cd + 1$ and $R_{00}(c,d) = 4cd + 2c + 1$.
- $Q_{11}(c,d) = 2cd + 2d - 1$ and $R_{11}(c,d) = 8cd + 8d - 2c - 8$.
- $Q_{01}(c,d) = 2cd + 1$ and $R_{01}(c,d) = 8cd + 2c + 2$.
- $Q_{10}(c,d) = 2cd + 2d - 1$ and $R_{10}(c,d) = 4cd + 2c + 4d - 1$.

Now we consider rationals whose even expansion has the form $\chi_1, ..., \chi_{2n}$, where each χ_j is a chunk. We break the even expansion into pairs of chunks, namely $(\chi_1, \chi_2)(\chi_3, \chi_4)....$ To each chunk χ_j we have a polynomials $Q(\chi_j)$ and $R(\chi_j)$. Given a pair of chunks, (χ_j, χ_{j+1}), we define

$$\Omega(\chi_j, \chi_{j+1}) = \frac{\log(R(\chi_j)R(\chi_{j+1}))}{\log(Q(\chi_j)Q(\chi_{j+1}))}. \tag{19.19}$$

LEMMA 19.8.

$$\Omega(\chi_j, \chi_{j+1}) \leq \frac{\log 72}{\log 9} \tag{19.20}$$

except possibly when all the numbers in the patches are less than 6.

Proof: We make explicit calculations for all pairs of chunks where the maximum term is less than 20. For the other cases, this is just a long exercise in calculus. For example, consider the case when χ_j has type $(0,0)$ and χ_{j+1} has type $(0,1)$. We are interested in the expression

$$f(c_1,d_1,c_2,d_2)=\frac{\log(4c_1d_1+2c_1+1)}{\log(2c_1d_1+1)}\oplus\frac{\log(8c_2d_2+2c_2+1)}{\log(2c_2d_2+1)}. \tag{19.21}$$

Here $\oplus$ denotes *Farey addition:*

$$\frac{u_1}{v_1}\oplus\frac{u_2}{v_2}=\frac{u_1+u_2}{v_1+v_2}. \tag{19.22}$$

When $\max(c_1,d_1)\geq 9$, each of the Farey summands in Equation 19.21 is easily seen to be less than $\log 72/\log 9$, and hence so is the Farey sum. The other cases are similar. □

LEMMA 19.9. *Suppose that all the numbers in the chunks χ_j and χ_{j+1} are less than 6. Then there are constants Q and R (depending on the case) such that $q_j\geq Qq_{j+2}$ and $r_j\leq Rr_{j+2}$ and $\log(Q)/\log(R)\leq\log(80)/\log(11)$.*

Proof: This is a computer calculation. The routines for the calculation are contained in the **Mathematica** directory in the source code for OctaPET. We work out one case of Lemma 19.9 by hand, to explain how the calculation works in general. Suppose $\chi_j=(2,2)$ and $\chi_{j+1}=(2,3,1)$. The signed C.F.E. corresponding to the sequence $(2,2,2,3,1)$ is $(2,1,2,\overline{e})$. This gives us the recurrence relations

- $q_1=2\theta_1+q_2$.
- $\theta_1=q_2+\theta_2$
- $q_2=2\theta_2+q_3$.
- $\theta_2=2q_3-\theta_3$.

This leads to $q_1=19q_3-8\theta_3\geq 11q_3$.

The recurrence relation for r_1 is

- $r_1=2\rho_1+r_2$ (Rule 1)
- $\rho_1=2r_2+\rho_2$ (Rule 1)
- $r_2=2\rho_2+\rho_3$ (Rule 1)
- $\rho_2=\rho_3+3r_3$ (Rule 5)
- $\rho_3=r_3+\rho_4$ (Rule 2)

This leads to $r_1=53r_3+17\rho_4\leq 70r_3$. □

Note that $\log(80)/\log(11)<\log(72)/\log(9)$. Hence, Lemmas 19.8 and 19.9 combine to give us the general estimate.

$$\frac{\log r_j}{\log p_j}\leq\frac{\log 72}{\log 9}\oplus\frac{\log r_{j+2}}{\log q_{j+2}}. \tag{19.23}$$

This fact combines with Equation 19.1 to finish the proof of Theorem 1.6.

Part 4

Topological Properties

Here is an overview of this part of the monograph:

- In §20 we prove several topological results which help us prove Statement 1 of Theorem 1.7. At the end of §20 we use these results to deduce some structural results about the unstable orbits for oddly even rational parameters.
- In §21 we combine the Shield Lemma and the Pinching Lemma to prove Statement 1 of Theorem 1.7, that Λ_s is a curve when s is oddly even.
- In §22 we prove some technical symmetry results which will help us prove Statements 2 and 3 of Theorem 1.7.
- In the short §23 we prove Statement 2 of Theorem 1.7, that Λ_s is a finite forest provided that $R^n(s)$ is eventually oddly even.
- In §24 we prove Statement 3 of Theorem 1.7, that Λ_s is a Cantor set when s is such that $R^n(s) > 1/2$ infinitely often. The proof is a bootstrap argument. The basic idea is that renormalization tends to make connected components of Λ_s larger, and eventually they get so large that we can rule them out.
- In §25 we prove Theorem 1.8, which explains the action of the PET f_s on the limit set Λ_s when s is oddly even. Along the way, we prove the analogous result for oddly even rational parameters.

CHAPTER 20

Controlling the Limit Set

20.1. The Shield Lemma

In this chapter we prove three results which help us control the topology of the limit set. The first result works just for oddly even parameters and the other two results work in general.

Let A_s be the symmetric piece from §9.

LEMMA 20.1 (Shield). *Let s be irrational and oddly even. Every point of ∂A_s, except the 2 vertices having obtuse angles, is contained in the edge of square. Those points which belong to the boundaries of more than one tile are the vertices of pairs of adjacent squares.*

Figure 20.1 illustrates the Shield Lemma. When s is rational, there are 2 small triangles touching the obtuse vertices of A_s. These triangles vanish in the irrational limit. The reason for the name of the lemma is that the structure shields Λ_s from the boundary of A_s, except at the two obtuse corners.

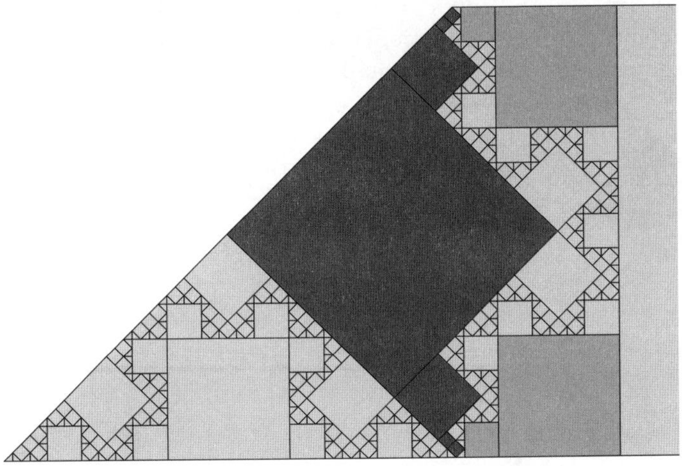

Figure 20.1: A_s for $s = 26/71$.

We define the *shield* Σ_s to be the union of the top left edge of A_s and the left half of the top edge. Σ_s is the union of two line segments. The top left vertex ν_s of X_s (and A_s) is the place where the two line segments join. By symmetry, it suffices to prove the Shield Lemma for the points of ∂A_s contained in the shield. We analyze the picture in the rational case and then take limits. In this section we work out how Δ_s sits in A_s when s is rational and oddly even.

We say that a square of Δ_s *abuts* Σ_s if an edge of the tile is contained in Σ_s. We call this segment the *contact* between the square and the shield. The *radius* of

T is the distance from the center of T to a corner of T. We call a radius ρ *realized*, if a square of Δ_s having radius ρ abuts the shield.

LEMMA 20.2. *Let s be an oddly even rational. The following is true:*

(1) *Exactly one triangle of Δ_s abuts Σ_s and the segment of contact contains ν_s as an endpoint.*
(2) *The squares which abut Σ_s occur in monotone decreasing size, largest to smallest, as one moves from an endpoint of Σ_s to ν_s.*
(3) *The number of squares of each size is determined by the even expansion of s.*
(4) *Let ρ be a realized radius. Some square of radius ρ, which abuts Σ_s, has a vertex within ρ of ν_s.*

Let us assume Lemma 20.2 for now, and finish the proof of the Shield Lemma. Let $\{r_n\}$ be a sequence of oddly even rationals which converges to s. Given the convergence of tilings described above, we see that the union of square tiles abutting Σ_s is the Hausdorff limit of the union of square tiles abutting Σ_{r_n}, as $n \to \infty$. The size of the single triangle in the picture for the rational parameters tends to 0. Hence, every point of $\Sigma_s - \nu_s$ is contained in a segment of contact for some square. The main point to worry about is that somehow there is a point $p \in \Sigma_s - \nu_s$, with the following property: As n tends to ∞, the square whose segment of contact contains p tends to 0 in size. This unfortunate situation cannot occur because it would violate Item 4 of Lemma 20.2.

Proof of Lemma 20.2: The proof goes by induction on the length of the orbit $\{R^n(s)\}$. When $s = 1/2$ or $s = 1/3$, the result holds by inspection. The case $s = 1/n$ follows from the Insertion Lemma. For $s \neq 1/n$, let $t = R(s)$. By induction, all the properties of the lemma hold for Δ_t.

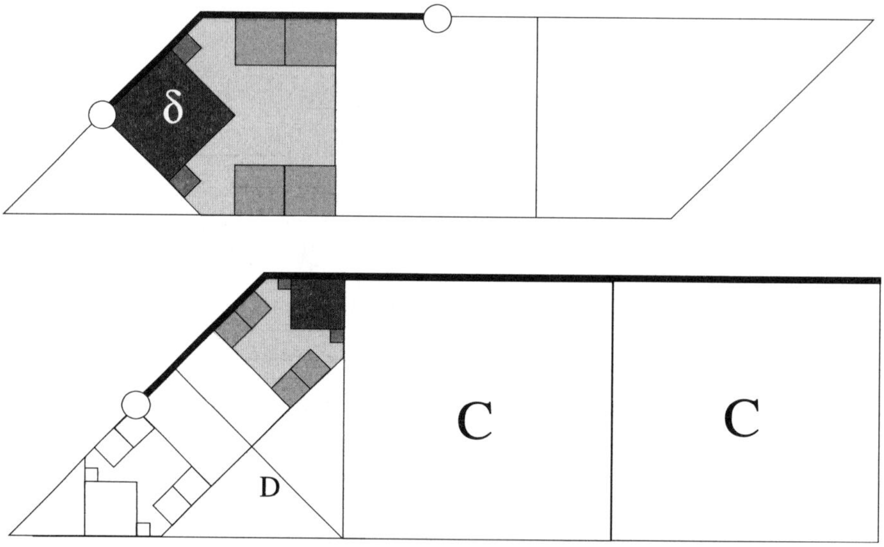

Figure 20.2: Inherited structure

Let $t = R(s)$. Let ϕ_s be the map from the Main Theorem. Let R_D denote the reflection in the diagonal line D_s of symmetry. The shading in Figure 20.2 is designed to help track the action of the map $f = R_D \circ \phi_s$. The top of Figure 20.2

shows some tiles of Δ_t and the bottom shows some tiles of Δ_s. The top shaded region is A_s and the shaded bottom region is $f(A_s)$. The diamond δ at the top left is the central tile of Ψ_t^0, the subset of Δ_t defined relative to the pair $(t, R(t))$. The box $f(\delta_t)$ abuts the leftmost central square of Δ_s. The pattern of tiles abutting the shield Σ_s is the same as the pattern of squares abutting the shield Σ_t, except that the the tiles marked C, which are central tiles of Δ_s, have been appended. The number of these extra tiles is determined by the first number n_0 in the even expansion of s. All the points in our lemma follow from this structure. □

COROLLARY 20.3. *Let S_s denote the left half of $\widehat{\Lambda}_s$. There exists a convex set D_2 such that $S_s \cap \text{interior}(A_s) \subset D_2$ and D_2 intersects ∂A_s only at the two obtuse vertices.*

Proof: We simply slice off from A_s suitably chosen neighborhoods of the square tiles which abut the edges of A_s. With a little care (i.e., by making these neighborhoods shrink very rapidly as we approach the vertices) we can make the resulting set convex. □

20.2. Another Version of the Shield Lemma

Here we prove a variant of the Shield Lemma. We say that two tiles σ and τ of Δ_s are *specially related* if σ and τ intersect along a segment, and both tiles have at least one an edge in ∂X_s. We say that a *special cycle* is a finite, cyclically ordered collection of tiles of Δ_s, such that every two consecutive ones are specially related. Figure 20.3 shows a special cycle of length 6.

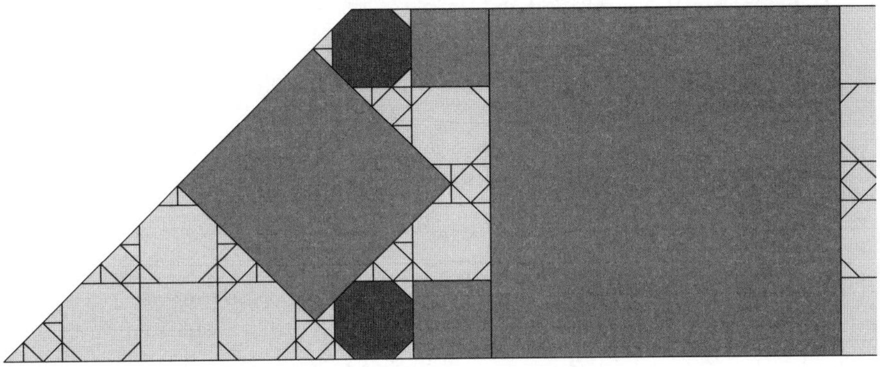

Figure 20.3: A special cycle for $s = 9/25$.

LEMMA 20.4 (Shield II). *Let $s < 1/2$ be any parameter with the property that $R^n(s) > 1/2$ for some $n > 0$. Then Δ_s has a special cycle which separates the horizontal edges of A_s from each other and from the diagonal edges of A_s.*

Proof: Let $p_{s,\pm}$ denote the upper and lower left vertices of the symmetric piece A_s, respectively. Note that $p_{s,+}$ is also the top left vertex of X_s^0. Let $p_{s,0}$ denote the leftmost vertex of A_s. When $s > 1/2$, the two points $p_{s,\pm}$ are vertices of O_s.

Consider first the case when $n = 1$. This means that $s < 1/2$ and $t = R(s) > 1/2$. In this case, by the Main Theorem, both $p_{s,+}$ and $p_{s,-}$ are vertices of $\phi_s(O_t)$. This tile abuts the leftmost central tile of Δ_s. In this case, the cycle has the form

$$\tau_1, ..., \tau_k, \phi_s(O_t), \tau_k, ..., \tau_2.$$

Here $\tau_1, ..., \tau_k$ are the central tiles of Δ_s, starting from the middle one and moving to the left. Figure 20.4 shows this situation for $s = 5/18$. Here $k = 1$ and the cycle has length 2. Evidently, the cycle has the claimed separation properties.

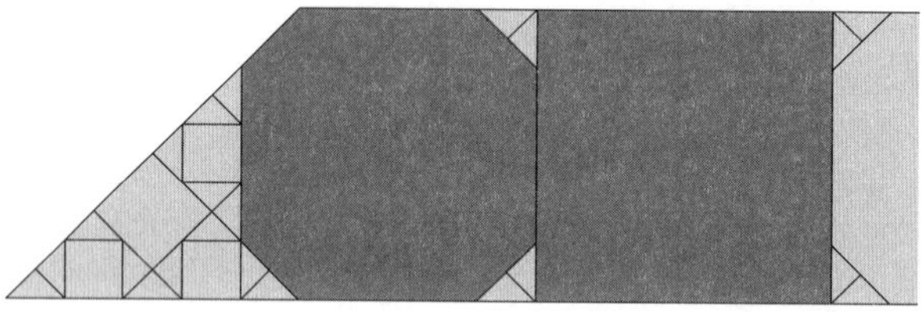

Figure 20.4: A special cycle for $s = 5/18$.

Suppose that $n > 1$, so that $s < 1/2$ and $t = R(s) < 1/2$. Let C_s and C_t respectively denote the middle central tiles of Δ_s and Δ_t. These are the two tiles on which the PET map is the identity. It follows from induction on n that $p_{s,\pm}$ is the vertex of an octagonal tile $O_{s,\pm}$ of Δ_s. By the Main Theorem, the tile $D_s = \phi_s(C_t)$ has $p_{s,0}$ as a vertex. Here we are choosing the left branch of the extension of ϕ_t in order to define D_t.

The same inductive argument as in the proof of Lemma 20.2 establishes two things at the same time.

- There is a finite sequence $C_s = \sigma_0, ..., \sigma_k = O_{s,+}$ of specially related tiles, all having edges in the top edge of X_s^0.
- There is a finite sequence $D_s = \tau_0, ..., \tau_\ell = O_{s,+}$ of specially related tiles, all having edges in the left edge of X_s^0.

The same statements hold with respect to $O_{s,-}$, by symmetry. When we concatenate the 4 sequences in the correct order, we get a special chain with the claimed separation properties. □

COROLLARY 20.5. Suppose that $s < 1/2$ is irrational and $R^n(s) > 1/2$ for some $n > 0$. Then a connected component of Λ_s cannot contain points on both horizontal edges of A_s, or points on a horizontal edge and a diagonal edge of A_s, or points on both horizontal edges of X_s^0. In particular, a connected component of Λ_s which contains a point in the interior of a bottom edge of A_s must lie entirely in A_s.

Proof: The only case that is not immediate is that statement that a connected subset of Λ_s cannot contain points on both horizontal edges of X_s^0. Such a subset would necessarily contain a smaller connected subset having points on the top edge of A_s and the diagonal edge of slope -1. □

20.3. The Pinching Lemma

This result in this section refers to the lines H, V, D_s, E_s considered in §9. (The lines H and V do not depend on s.) In order to save words, we will talk about the *lines* of symmetry when we really mean to speak about the line segments of symmetry which are contained inside the set X_s^0.

LEMMA 20.6 (Pinching). *At most one point of* Λ_s^0 *lies on each line of symmetry.*

We will assume the Pinching Lemma is false and derive a contradiction. The argument we give here is a prototype for similar arguments we give several times in later chapters.

Say that a *counterexample* is a quadruple $\Theta = (L, p_1, p_2, s)$, where $L = L_s$ is a fundamental line of symmetry and $p_1 \neq p_2 \in L \cap \Lambda_s^0$. We call s the *parameter* of the counterexample. We define

$$\lambda(\Theta) = \frac{\|p_1 - p_2\|}{\text{diam}(X_s^0)} \tag{20.1}$$

We let M denote the supremum over all counterexamples.

Let Z_s be the set from the Main Theorem. Let Z_s^0 be the left half of Z_s. We define

$$\heartsuit_s = \frac{\text{diam}(X_s^0)}{\text{diam}(Z_s^0)} \tag{20.2}$$

(1) $\heartsuit_s > 1$ for all $s \in (0, 1)$.
(2) $\heartsuit_s$ is uniformly bounded away from 1 when $s < 1/2$.

The second property follows from compactness. The point is that $\heartsuit_s$ does not tend to 1 as $s \to 1/2$.

The strategy of our proof is to start with a counterexample Θ_s, associated to the parameter s, and then produce a new counterexample Θ_t associated to $t = R(s)$, such that

$$\lambda(\Theta_t) = \heartsuit_s \lambda(\Theta_s).$$

When $s < 1/2$ this leads to a contradiction if we take $\lambda(\Theta_s) > M/\heartsuit_s$. When $s > 1/2$ this reduces us back to the case $s < 1/2$.

20.3.1. *Case 1.* Suppose that Θ_s is a counterexample and $s < 1/2$ and $t = R(s) > 1/2$. By the Insertion Lemma, it suffices to consider $s \in (1/4, 1/3)$.

Let $O_s = \phi_s(C_t)$, where C_t is the central tile of Δ_t. See Figure 20.5.

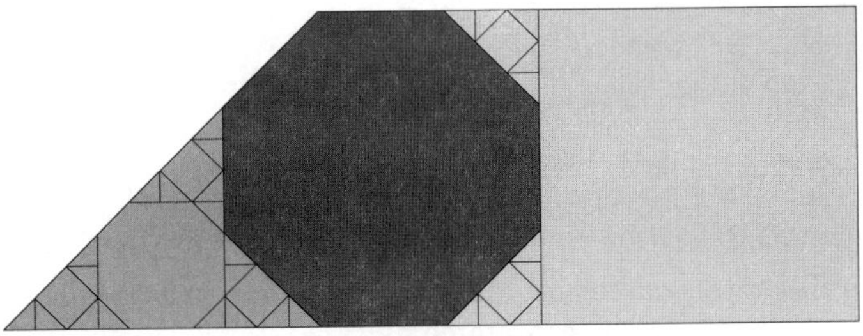

Figure 20.5: O_s (dark) and Z_s^0 (lightly shaded) for $s = 5/17$.

We observe the following:

- $E_s \subset Z_s^0$
- $V \subset Z_s^0$

- $H - O_s \subset Z_s^0$
- $R_D R_H(D_s - O_s) \subset Z_s^0$.

Suppose first that Θ_s involves one of E_s, V, H. Because O_s separates Z_s^0 from the rest of X_s^0, we have $\Theta_s \subset \Lambda_s \sqcap Z_s$, and the intersections of Θ_s with the relevant line of symmetry must occur on one of the listed segments. Hence

$$\Theta_t = \phi_s^{-1}(\Theta_s) \subset \Lambda_t. \tag{20.3}$$

The map ϕ_s^{-1} carries each $E_s, V, H - O_s$ to the lines of symmetry V, D_t, E_t. Hence, Θ_t is a counterexample associated to the parameter t.

Since $Y_t^0 = X_t^0$ for $t > 1/2$, we have

$$\lambda(\Theta_t) = \frac{\|\phi_s^{-1}(p) - \phi_s^{-1}(q)\|}{\text{diam}(Y_t^0)} = \frac{\|p - q\|}{\text{diam}(\phi_s(Y_t^0))} =$$

$$\frac{\|p - q\|}{\text{diam}(Z_s^0)} = \lambda(\Theta_s)\frac{\text{diam}(X_s^0)}{\text{diam}(Z_s^0)} = \heartsuit_s \lambda(\Theta_t). \tag{20.4}$$

This gives the contradiction mentioned above.

If Θ_s involves the line D_s, we let

$$\Theta_t = \phi_s^{-1} R_D R_H(\Theta_s) \subset \Lambda_t. \tag{20.5}$$

This time, Θ_t is a counterexample with respect to the parameter t and the line H. Now we proceed as above.

20.3.2. *Case 2.* Suppose that Θ_s is a counterexample and $s < 1/2$ and $t = R(s) < 1/2$. This puts $s \in (1/3, 1/2)$.

The argument is very similar to what we just did. This time, we let O_s be the pyramid from Lemma 15.8. The region O_s is darkly shaded in Figure 20.6. The lightly shaded regions show the locatons of the 4 lines of symmetry.

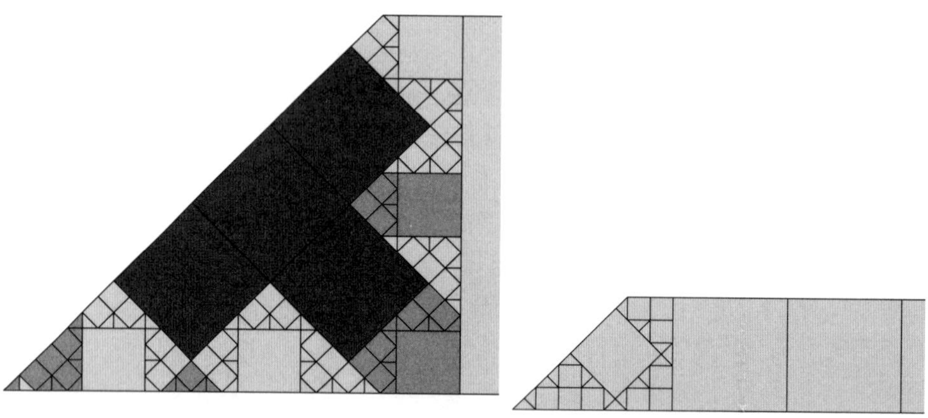

Figure 20.6: Δ_s for $s = 12/29$ and Δ_t for $t = R(s) = 5/24$.

Referring to the notation of the Filling Lemma, and working counterclockwise around the pyramid, we have the containments

- $E_s \subset Z_s^0$.
- $V - O_s \subset Z_s^j$.
- $D_s - O_s \subset R_D R_H \phi_s(A_t)$.
- $H - O_s \subset R_D Z_s^k$.

Here $j, k \in \{1, ..., \mho - 1\}$ are suitably chosen indices.

We can pull back by the relevant map, as in Case 1, and we get the same contradiction.

20.3.3. *Case 3.* Suppose that Θ_s is a counterexample and $s > 1/2$. By the Inversion Lemma, is suffices to take $s \in (1/2, \sqrt{2}/2]$. Referring to the Covering Lemma, this puts is in the case of Lemma 14.4.

We have the following containments:

- $D_s \subset Z_s$.
- $E_s \subset Z_s$.
- $V \subset Z_s$.
- $H \subset R_D\phi_s(A_t)$.

When $s > 1/2$, the map ϕ_s is an isometry. When we pull back by the relevant map, we get a counterexample Θ_t such that $\lambda(\Theta_t) = \heartsuit_s\lambda(\Theta_s)$. This does not give an immediate contradiction, but we still have $\lambda(\Theta_t) > \lambda(\Theta_s)$. So, we replace Θ_s by Θ_t and then we are back in Case 1 or Case 2.

This completes the proof of the Pinching Lemma.

20.4. Rational Oddly Even Parameters

Let s be an oddly even parameter. Let Ξ_s denote the union of the unstable tiles. Let L_s denote the union of the long sides of each triangle in Ξ_s. We call each individual triangle edge a *segment* of L_s. As usual L_s^0 denotes the left half of L_s, etc.

As a warmup to our proof of Statement 1 of Theorem 1.7, we consider what happens for oddly even rational parameters. Figure 20.7 shows Ξ_s^0 for two parameters. The goal of this section is to establish in general some of the features of L_s which are suggested by these pictures.

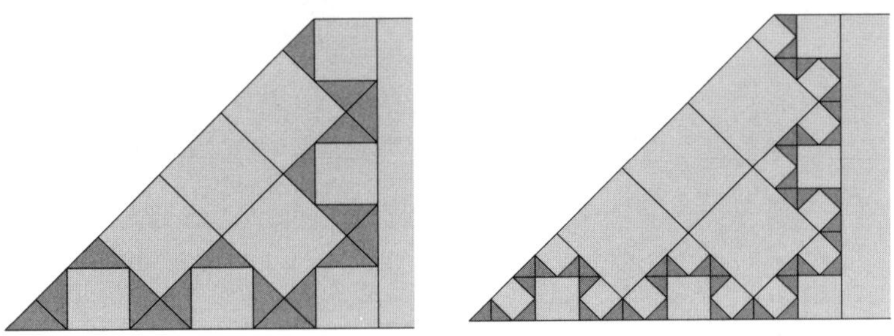

Figure 20.7: Ξ_s^0 for $s = 5/12$ and $s = 7/17$.

LEMMA 20.7. *Ξ_s^0 intersects a line of symmetry either in a single vertex, or in a short edge common to two adjacent triangles, or in an edge of symmetry of one of the triangles. Hence L_s^0 intersects a line of symmetry either in a single point. This point is either an endpoint or the midpoint of a segment. When L_s^0 meets a line of symmetry in a vertex, the segments of L_s^0 incident to this vertex meet the line of symmetry at an angle of $\pi/4$.*

Proof: We check the case $s = 1/2$ directly and then the case $s = 1/2n$ follows by the Insertion Lemma. The general case follows from induction, exactly along the lines that we proved the Pinching Lemma. $\square$

LEMMA 20.8. *L_s^0 intersects the shield in a single point, namely the top left vertex of X_s. The segment of L_s^0 incident to this vertex is perpendicular to the segment of L_s^0 incidendent to the bottom left corner of X_s.*

Proof: The first statement is an immediate corollary of Lemma 20.2. The second statement holds for $s = 1/2$ by inspection and then for $s = 1/2n$ by the Insertion. The general case follows from the Main Theorem and induction on the length of the even expansion of s. □

LEMMA 20.9. *Let s be oddly even and rational. L_s^0 is a polygonal arc connecting the two left vertices of X_s. The first and last sides of L_s^0 are perpendicular, and L_s^0 makes a right-angled turn after each segment.*

Proof: We check the result for $s = 1/2$ directly, and then the result follows for the case $s = 1/2n$ by the Insertion Lemma. The rest of the proof goes by induction on the length of the even expansion of s. Let $\mho$ be the layering constant for s. It follows from the Main Theorem and induction that $L_s \cap \Psi_s^0$ is a polygonal arc connecting the two bottom vertices of Ψ_s^0. By the Filling Lemma,

$$L_s \cap \Psi_s^j = T^j(L_s \cap \Psi_s^0), \qquad j = 0, ..., (\mho - 1). \tag{20.6}$$

Here T is the map from Equation 13.3. By construction, the two arcs $L_s \cap \Psi_s^i$ and $L_s \cap \Psi_s^j$ are disjoint when $|i - j| \geq 2$. By Lemma 20.8, these arcs meet at a single point when $|i - j| = 1$. Finally, the last segment of $L_s \cap \Psi_s^i$ meets the first segment of $L_s \cap \Psi_s^{i+1}$ at a right angle, by induction and Lemma 20.8. Hence, the whole polygonal arc makes a right-angled turn after each segment.

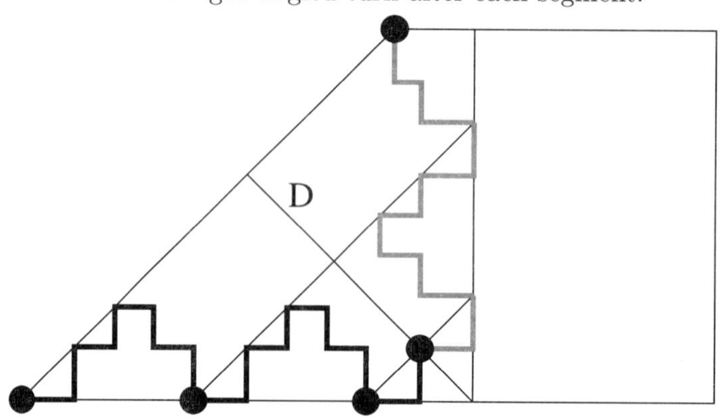

Figure 20.8: concatenation of arcs. Here $\mho = 2$.

Consider the subset α of $L_s \cap \Psi_s^{\mho}$ which lies beneath the diagonal line D of symmetry. This arc is the initial portion of $T^{\mho}(L_s \cap \Psi_s^0)$. Thus, α is an arc which connects the bottom right vertex of $\Psi_s^{\mho}$ to D. But L_s only intersects D once, by Lemma 20.8. Hence, the subset of L_s beneath D is a polygonal arc connecting the bottom left vertex of X_s to a point of D.

The reflection R_D maps the subset of L_s above D to a subset of the portion of L_s^0 below D. Hence, some initial segment of L_s^0 above D connects a point of D to the upper left vertex of X_s. But then Lemma 20.8 implies that this initial segment is all of L_s^0 above D. □

CHAPTER 21

The Arc Case

21.1. The Easy Direction

Let $s \in (0,1)$ be irrational. Our goal is to prove that Λ_s is a disjoint union of two arcs if and only if s is oddly even. By symmetry, the result we want is equivalent to the statement that Λ_s^0 is an arc if and only if s is oddly even.

LEMMA 21.1. *For each integer k such that $R^k(s) > 1/2$, there is an octagon O_k having one edge in the left side of X_s and one edge in the bottom side of X_s. If there are two istinct indices k and ℓ with this property, then the octagons O_k and O_ℓ are distinct.*

Proof: Say that an octagon is *wedged* in a parallelogram if one edge of the octagon lies in the left edge of the parallelogram and another edge lies in the bottom edge, as in Figure 21.1.

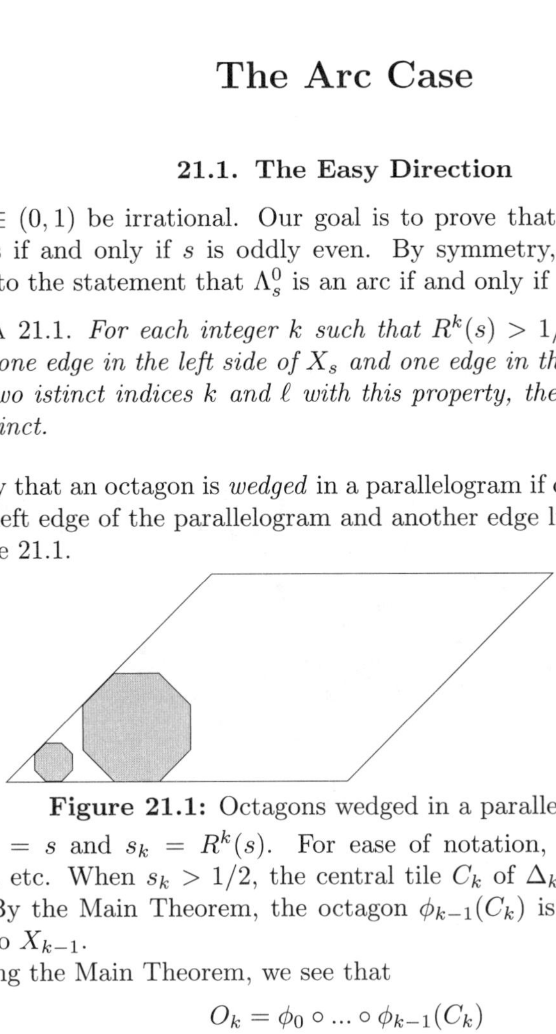

Figure 21.1: Octagons wedged in a parallelogram.

Let $s_0 = s$ and $s_k = R^k(s)$. For ease of notation, we set $\phi_k = \phi_{s_k}$ and $\Delta_k = \Delta_{s_k}$, etc. When $s_k > 1/2$, the central tile C_k of Δ_k is an octagon wedged into X_k. By the Main Theorem, the octagon $\phi_{k-1}(C_k)$ is a tile of Δ^0_{k-1}, and is wedged into X_{k-1}.

Iterating the Main Theorem, we see that

$$O_k = \phi_0 \circ \ldots \circ \phi_{k-1}(C_k) \tag{21.1}$$

is wedged into X_0.

Suppose that $\ell > k$ is another index such that $s_\ell > 1/2$. Then the two octagons C_k and $\phi_k \circ \ldots \circ \phi_{\ell-1}(C_\ell)$ are distinct because one octagon is the central tile of Δ_k and the other one is not. But then O_k and O_ℓ are the images of the above octagons under the same similarity. Hence, they are distinct. □

COROLLARY 21.2. If $R^n(s) > 1/2$ for at least K different positive indices, then Λ_s^0 has at least $K+1$ connected components.

Proof: The K octagons guaranteed by Lemma 21.1 are all distinct. Call these octagons $O_1, \ldots, O_K$. Each of these octagons is wedged into X_s, and so the union of these octagons separates X_s^0 into $K+1$ connected components. We just need to see that Λ_s^0 intersects each component.

Each octagon O_j has two vertices in the bottom edge of X_s. At each of these vertices, the adjacent edge of O_j makes an acute angle with the bottom edge of X_s. (The angle is $\pi/4$.) Since the Δ_s consists of an open dense (in fact full measure) set of squares and semi-regular octagons, every neighborhood of the two vertices in question must intersect infinitely many tiles of Δ_s. Hence, the two bottom vertices of O_j lie in Λ_s^0. This proves what we want. □

What we have shown is that Λ_s^0 is not an arc if $R^n(s) > 1/2$ for some $n > 0$.

LEMMA 21.3. *If $s > 1/2$, then Λ_s^0 is not an arc.*

Proof: When $s > 1/2$ the region $\boldsymbol{X}_s^0$, the left side of the central octagon, is a kite-shaped region with 3 corners having angle $\pi/4$ radians. By the same reasoning as as in Corollary 21.2, all three corners must belong to Λ_s^0. But then Λ_s^0 cannot be an arc. □

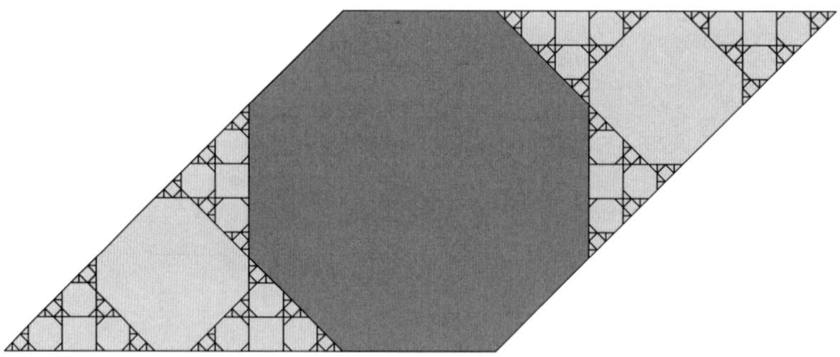

Figure 21.2: Δ_s for $s = 20/29$.

We now know that Λ_s^0 is an arc only if s is oddly even. The rest of the chapter is devoted to proving that Λ_s^0 is an arc when s is oddly even.

21.2. A Criterion for Arcs

To show that Λ_s^0 is an arc when s is oddly even, we will use the topological criterion we establish in this section. Some version of this criterion is certainly well known. Lacking a reference which states precisely what we want, we prove the result here.

Say that a *marked piece* is a compact, embedded, convex set with two distinguished vertices. Say that a *chain* is a finite union $D_1, ..., D_n$ of marked pieces such that $D_i \cap D_{i+1}$ is one point, and that this point is one of the marked points on each of D_i and D_{i+1}. We also require that $D_i \cap D_j = \emptyset$ for all other indices $i \neq j$. We define the *mesh* of the chain to be the maximum diameter of one of the marked pieces.

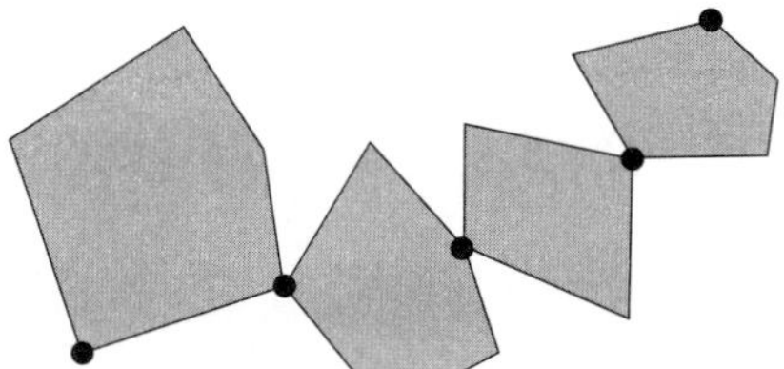

Figure 21.3: A chain of length 4.

We say that a compact set S *fills* a chain $D_1, ..., D_n$ if $S \subset \bigcup D_i$ and S contains every marked point of the chain. The purpose of this section is to establish the following criterion.

LEMMA 21.4 (Arc Criterion). *Let S be compact. Suppose, for every $\epsilon > 0$, that S fills a chain having mesh less than ϵ. Then S is an embedded arc.*

We will assume that S satisfies the hypotheses of the lemma, and then show that S is an arc.

LEMMA 21.5. *S is connected.*

Proof: If S is disconnected, then we can write $S = S_1 \cup S_2$ where S_1 and S_2 are separated by some $d > 0$. But then S could not fill a chain of mesh size less than d. □

LEMMA 21.6. *Suppose that S fills a chain $C_1, ..., C_m$. Then there is some $\epsilon > 0$ with the following property. If S also fills a chain $D_1, ..., D_n$ having mesh size less than ϵ, then S fills a chain $E_1, ..., E_p$ where each E_i has the form $C_j \cap D_k$.*

Proof: We can choose ϵ so small that each D_j has the following properties:

- The diameter of D_j is smaller than the length of any edge of any C_i.
- D_j intersects any C_i in at most 2 edges.
- D_j cannot intersect C_i and C_k if i and k are not consecutive indices.

For each i, there are unique and distinct pieces D_{j_1} and D_{j_2} which contain the two marked points of C_i.

We claim that the pieces between D_{j_1} and D_{j_2} have both marked points inside C_i. If our claim was false, then some D_j, with $j_1 < j < j_2$, would have one marked point in C_i. Note that D_j must have another marked point in either C_{i-1} or C_{i+1}, because this marked point is a vertex of S. Suppose that D_j has its other marked point in C_{i+1}. Then $D_j \cap (C_i \cup C_{i+1})$ is disconnected because D_j does not contain the vertex $C_i \cap C_{i+1}$. Hence

$$D_1 \cup ... \cup D_{j-1} \cup \big(D_j \cap (C_i \cup C_{i+1})\big) \cup D_{j+1} \cup ... \cup D_n$$

consists of two disconnected components, each of which intersects S nontrivially. This contradicts the connectivity of S.

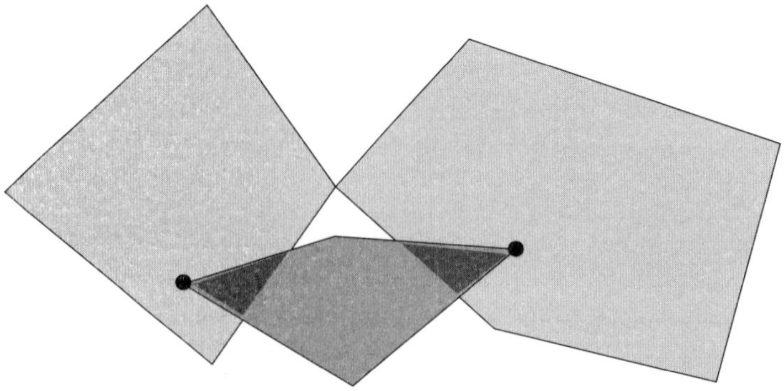

Figure 21.4: Disconnected set.

Now that we know the claim is true, we form the portion of the E-chain inside C_i by taking the intersections $C_i \cap D_j$ and using the marked points of D_j for $j_1 < j < j_2$. See Figure 21.4 below. The marked points of $C_i \cap D_{j_1}$ are the marked points of C_i inside of D_{j_1} and the marked point of D_{j_1} inside C_i. Similarly for $C_i \cap D_{j_2}$. We do the same thing for each i and this gives us the conclusion of the Lemma. □

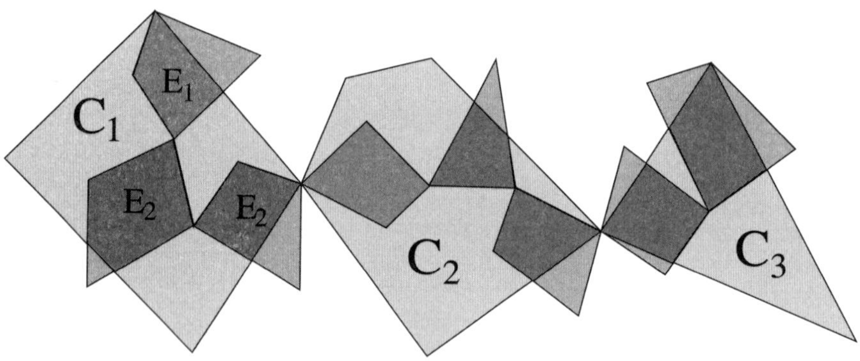

Figure 21.5: Refining a chain.

If the chain $C_1, ..., C_m$ and the chain $E_1, ..., E_p$ are related as in the previous lemma, we say that $E_1, ..., E_p$ *refines* $C_1, ..., C_m$. In view of the previous result, we can assume that S fills an infinite sequence $\{\Omega_i\}$ of chains such that each one refines the previous one and the mesh size tends to 0.

For each i, we inductively create a partition P_i of $[0, 1]$ into intervals, such that the number of intervals coincides with the number of marked pieces in Ω_i, in the following manner. Once P_i is created, we distribute the intervals of P_{i+1} according to how Ω_i contains Ω_{i+1}. If the kth piece of Ω_i contains n_k pieces of Ω_{i+1}, then P_{i+1} is created from P_i by subdividing the kth interval of P_i into n_k intervals of equal size. Note that the mesh size of P_i tends to 0 as i tends to ∞.

There is a bijective correspondence between marked pieces in the chains and intervals in the partition. The correspondence respects the containment and intersection properties. For instance, two marked pieces intersect if and only if the corresponding intervals share an endpoint. Each point of S is contained in an infinite nested intersection of marked pieces, and we map this point to the corresponding nested intersection of intervals. This map is clearly a homeomorphism. The inverse map gives a parameterization of S as an arc in the plane.

21.3. Elementary Properties of the Limit Set

Let A_s be the symmetric piece from §9.

LEMMA 21.7. *Λ_s^0 contains the two left vertices of X_s and the two obtuse vertices of A_s.*

Proof: Let v be one of left vertices of X_s. Since the angle of X_s at v is not a right angle, there must be infinitely many squares contained in every neighborhood of v.

Note that the top left vertex of X_s^0 is also the top obtuse vertex of A_s. So, Λ_s^0 contains the top obtuse vertex of A_s. By symmetry, Λ_s^0 contains the bottom obtuse vertex of A_s. □

We know from Statement 3 of Corollary 1.3 that Δ_s consists entirely of squares. These squares are necessarily diamonds or boxes.

LEMMA 21.8. *Suppose $\gamma \subset X_s^0$ is a compact arc which connects a point in a box to a point in a diamond. Then γ contains a point of Λ_s^0.*

Proof: Compare our proof of Statement 2 of Theorem 1.5. By compactness, it suffices to show that arbitrarily small perturbations of γ contain points of Λ_s^0. Hence, we may perturb so that γ does not contain any vertices of any tiles in Δ_s. Suppose γ does not intersect Λ. Then γ only intersects finitely many tiles, $\tau_1, ..., \tau_n$. Moreover, τ_i and τ_{i+1} must share an edge. Hence, by induction, τ_1 is a box if and only if τ_n is a box. But τ_1 is a box and τ_n is a diamond. This is a contradiction. □

LEMMA 21.9. *Each fundamental line of symmetry contains a point of Λ_s^0.*

Proof: To make the argument cleaner, we attach a large diamond δ to the picture along the left edge of X_s, and we attach a large box β to the picture along the bottom edge of X_s. These extra squares are disjoint from X_s except along the relevant edges. Once we add these two squares, we see that each of the lines in question connects a diamond to a box. H connects δ to a the leftmost central tile of Δ_s and both V, D, E all connect β to δ. By Lemma 21.8, each of these lines contains a point of Λ_s^0. □

21.4. Verifying the Arc Criterion

Our strategy is to show that Λ_s^0 satisfies the Arc Criterion established in §21.2. Suppose that Λ_s^0 fills some chain $D_1, ..., D_n$. We call this chain *good* if:

- D_j is disjoint from the interiors of the edges of ∂A_s, for $j = 1, ..., n$.
- The first marked point of D_1 is the bottom left vertex of X_s.
- The last marked point of D_n is the top left vertex of X_s.

Here A_s is the symmetric piece, as above. Some of the D_j will lie in A_s and the rest will meet A_s in at most 1 point.

LEMMA 21.10. *Λ_s^0 fills a good chain.*

Proof: Our chain has two pieces. We set $D_1 = B_s$, the triangle from §9, and we let $D_2 \subset A_s$ be the set from Corollary 20.3. □

LEMMA 21.11. *Let $t = R(s)$. Suppose S_t fills a good chain having mesh m. Then Λ_s^0 fills a good chain having mesh at most $m/\sqrt{2}$.*

Proof: Let C_t be the good chain filled by S_t. We use the notation from the Filling Lemma, Equation 13.3, and the Main Theorem. We make our construction in 5 steps. Let $\mho$ be the layering constant for s.

Step 1: For $j = 0, ..., \mho$, we define $C_j = T_s^j \circ \phi_s(C_t)$. Figure 21.6 illustrates our construction. The individual chains $C_0, ..., C_{\mho-1}$ piece together to make one long chain because the second disk of C_j touches the common edge between Ψ_s^j and Ψ_s^{j+1} only at the bottom vertex of this edge.

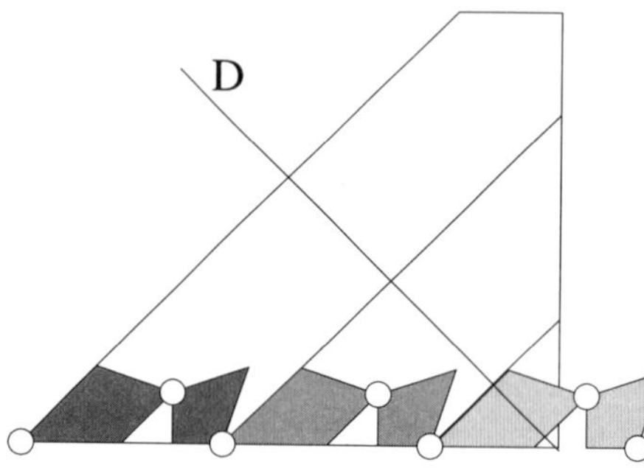

Figure 21.6: Step 1: The chains C_j for $j = 0, ..., \mho$. Here $\mho = 2$.

Step 2: The problem with $C_\mho$ is that some of it sticks over the edge of X_s^0. This is the lightly shaded set in Figure 21.6. However, we know from the Pinching Lemma that Λ_s^0 intersects the symmetry line D_s in a single point. All other points of D_s must have neighborhoods contained in finitely many squares. For this reason, we can make the essentially the same construction as in Corollary 20.3 to produce a convex disk $U \subset \Psi_s^\mho$ such that $\Lambda_s^0 \cap \Psi_s^\mho \subset U$ and $U \cap D_s$ is the single point which belongs to Λ_s^0. See Figure 21.7.

We now improve $C_\mho$ as follows. We intersect each piece of $C_\mho$ with the set U and throw out all those after the first one which is disjoint from U.

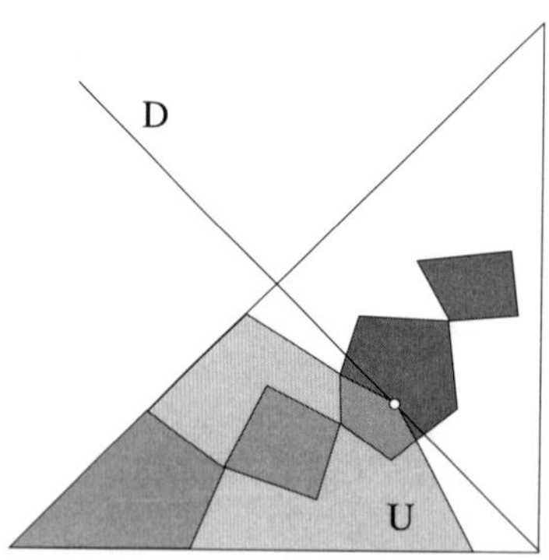

Figure 21.7: Step 2: Improving $C_\mho$.

The result is a chain which joins the bottom vertex of the edge $\Psi_s^{\mho-1} \cap \Psi_s^k$ to the point $\Lambda_s^0 \cap D_s$. Figure 21.7 shows the construction. We call this improved chain $C'_\mho$. Define $\Upsilon_0 = C_0, ..., C_{\mho-1}, C'_\mho$. By construction, this chain is filled by the portion of Λ_s^0 beneath the line D_s.

Step 3: Define $\Upsilon_2 = \Upsilon_0, \Upsilon_1$ where $\Upsilon_1 = R_D(\Upsilon_0)$ is the reflected chain. That is, we continue our chain by reflecting it across the line D. The resulting chain contains Λ_s^0, by symmetry, but we are not quite done.

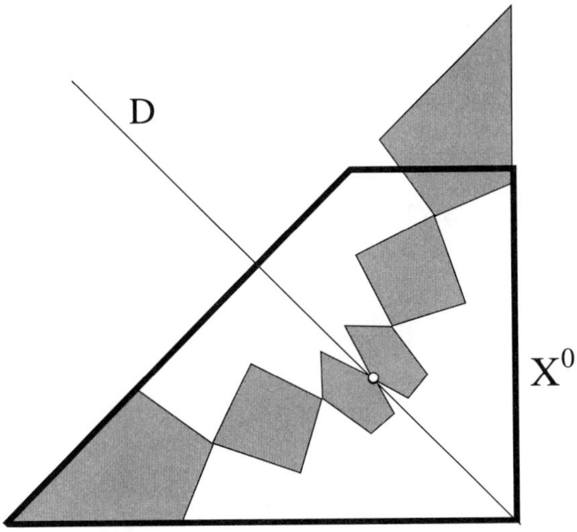

Figure 21.8: Step 3: Extending by reflection

Some of the final pieces of Υ_1 might not lie in X_s. The problem is that X_s is not symmetric with respect to R_D. The portion below D is larger than the portion above D.

Step 4: We finish the construction by a method very similar to what we did in Step 2. We simply intersect all the pieces of Υ_1 with X_s^0, and let Υ_1' and omit all those pieces which come after the first one which has trivial intersection with X_s^0. We set $\Upsilon_3 = \Upsilon_0, \Upsilon_1'$. By construction, Υ_3 is a chain filled by Λ_s^0. Moreover, since ϕ_s contracts distances by some $\lambda_s < 1/\sqrt{2}$. we see that the mesh of Υ_3 is less than $m/\sqrt{2}$.

Step 5: The chain Υ_3 might not satisfy the first goodness condition. To remedy this, we shrink the pieces slightly (away from the marked points) so that they are all disjoint from the interiors of the edges of A_s. What allows us to do this is the Shield Lemma combined with compactness. The final chain has all the desired properties. □

Note that if s is oddly even, then so is $R(s)$. The chains in Lemma 21.10 all have mesh size less than 2. It now follows from iterating Lemma 21.11 that Λ_s^0 fills a good chain having mesh size less than ϵ, for any given ϵ_0. Our Arc Criterion how shows that Λ_s^0 is an arc. This completes the proof.

CHAPTER 22

Further Symmetries of the Tiling

22.1. Zones

Our constructions refer to the pyramids defined in §15.3. We define a *zone* to be a right-angled isosceles triangle ζ having the following structure:

- The long side of ζ lies in the bottom edge of X_s^0.
- The apex of ζ is either a center or a corner of a square in Ψ_s.
- ζ lies beneath the line Z_s of bilateral symmetry.

Figure 22.1 shows an example.

Figure 22.1: A zone (dark) of the octagrid, for $s = 7/17$.

The zone ζ shown in Figure 22.1 is a clean set: The boundary $\partial\zeta$ does not intersect any tile interiors. Note that ζ need not be a clean set in general. In general, we define $\Delta_s \cap \zeta$ to be the union of tiles which have an interior point in ζ. Figure 22.2 shows an example. In this example, the right edge of ζ is D_s and the apex of ζ is the center of the dark square in Ψ_s.

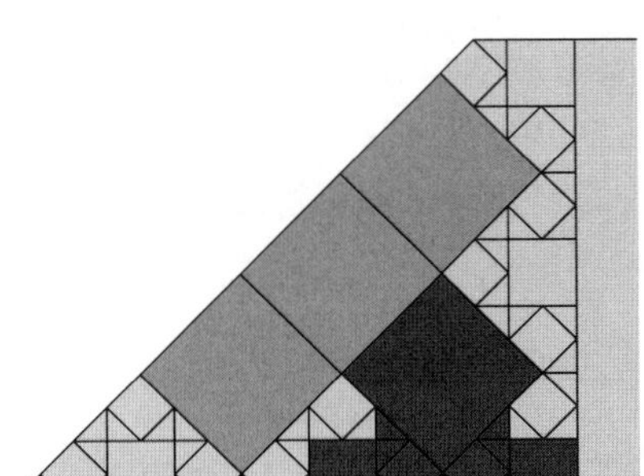

Figure 22.2: $\Delta_s \cap \zeta$ (dark) for $s = 7/17$.

22.2. Symmetry of Zones

We think of the zones as playing a role similar to the role played by symmetric pieces. It appears that the zones always have bilateral symmetry, and this is what we will prove.

LEMMA 22.1. *Let Γ_v be a zone. Then reflection in the vertical axis of Γ_v is a symmetry of $\Delta_s \cap \Gamma_v$.*

Proof: Let V, H, D_s be the vertical, horizontal, and diagonal lines of symmetry from §9. Let B_s, A_s, P_s be the corresponding symmetric pieces. We consider two special cases first. Suppose first that $v \in V$. In this case $\Gamma_v \subset B_s$ and R_V is the bilateral symmetry of $\Gamma_v \cap \Delta_s$. If $v \in R_D(H)$, then $\Gamma_v \subset P_s$, and H is the centerline of $R_D(\Gamma_v)$, and $R_D(\Gamma_v) \subset A_s$. But then $R_D(\Gamma_v) \cap \Delta_s$ has bilateral symmetry. But then so does $\Gamma_v \cap \Delta_s$.

If v is the corner of a square in the pyramid, as in Figure 22.1, we have

$$\Gamma_v \subset \bigcup_{k=0}^{\mho} \Psi_s^k, \tag{22.1}$$

and there is a horizontal translation symmetry T^j from Equation 13.3 so that $T^j(v) \in V$. This map carries $\Gamma_v \cap \Delta_s$ to $T^j(\Gamma_v) \cap \Delta_s$, and we reduce to the first case we considered.

If v is the center of a square in the pyramid, then the same argument as above works, except when $v \in D_s$. Figure 22.2 shows an example of an exceptional case. In this case, we have

- $\Gamma_v \subset P_s$.
- $R_D(\Gamma_v) \subset A_s$.
- $R_H R_D(\Gamma_v) \subset P_s$.

The new zone

$$R_D R_H R_D(\Gamma_v)$$

is one of the zones already considered, and therefore intersects Δ_s in a set of bilateral symmetry. Hence, the same goes for Γ_v. $\square$

22.3. Intersections with Zones

To further the analogy between zones and symmetric pieces, we prove an analogue of the Pinching Lemma. Let $\Lambda_s \sqcap \zeta$ denote those points $p \in \zeta$ such that every open neighborhood of p contains infinitely many tiles of $\Delta \cap \zeta$. We say that a *zone centerline* is the vertical line of symmetry the zone.

LEMMA 22.2. *Let ζ be a zone. $\Lambda_s \sqcap \zeta$ intersects the centerline of ζ in at most one point.*

Proof: Let ζ be a zone. Let L be the centerline of ζ. The proof of Lemma 22.1 shows that we can find some isometry I such that $I(\zeta \cap \Delta) \subset \Delta$ and $I(L) \subset V$, the vertical line of symmetry. By the Pinching Lemma, Λ_s intersects V in at most one point. Hence $\Lambda_s \sqcap I(\zeta)$ is at most one point. But then

$$I^{-1}(\Lambda_s \sqcap I(\zeta)) = \Lambda_s \sqcap \zeta. \tag{22.2}$$

Hence $\Lambda_s \sqcap \zeta$ intersects L at most once. $\square$

COROLLARY 22.3. Suppose that γ is an embedded loop in Λ_s which lies beneath the line D_s of symmetry. Then L_s either lies to the left of the leftmost zone centerline or else lies between two consecutive zone centerlines.

Proof: The portion of Δ_s beneath D_s is a union of tiles in the pyramid and tiles in the zones. If γ does not have the properties advertised in the lemma, then γ intersects the centerline L of some zone ζ twice. Every point of L except the apex lies in the interior of ζ. Moreover, the apex of L is either the middle of a square tile or the corner of a square tile. In the corner case, the only time γ can intersect the apex is when the apex is on the bottom of the pyramid, as in Figure 22.1. So, in all case, if γ intersects L twice, so does $\Lambda \sqcap \zeta$. This contradicts our previous result. $\square$

Now we deduce a result which we will use when proving Statement 2 of Theorem 1.7. Let Z_s be the set from the Main Theorem.

LEMMA 22.4. *Suppose $s < 1/2$. Suppose that γ is an embedded loop in Λ_s. Then there is some other embedded loop γ', isometric to γ, contained in $\Lambda_s \sqcap Z_s$.*

Proof: By the Pinching Lemma, γ can only intersect the diagonal line D_s once. Since γ is an embedded loop, this means that γ either lies above D_s or below D_s. If γ lies above D_s then $\gamma \subset P_s$, the symmetric piece for which D_s is the centerline. Replacing γ by $R_D(\gamma)$ if necessary, we can assume that γ lies below D_s.

First consider the case when $t = R(s) > 1/2$. Let O_t be the central octagonal tile of Δ_t. As is illustrated in Figure 20.5, the tile $\phi_t(O_t)$ separates X_s^0 into 3 disconnected regions, the biggest of which is Z_s^0. If γ does not already lie in Z_s^0, then the isometric loop

$$\gamma' = R_D R_H R_D(\gamma) \tag{22.3}$$

lies in Z_s^0. This takes care of the case when $t > 1/2$.

Now consider the case that $t < 1/2$. Now we can use the results about zones. It now follows from the previous result that γ lies between two consecutive zone centerlines. If γ does not lie to the left of the leftmost zone centerline, then there is some zone ζ such that $\gamma \subset \zeta$ and γ lies to the right of the centerline L of ζ. Here we are crucially using the fact that the region between two zone centerlines and beneath the pyramid lies in a single zone.

Let $\rho : \zeta \to \zeta$ be reflection in L. The new loop $\rho(\gamma)$ still lies in Λ_s and moreover lies to the left of L. Repeating this reflection trick finitely many times, we finally produce an isometric loop which lies to the left of the leftmost zone centerline L_0. Here L_0 contains the left corner of the bottom left square in the pyramid, as shown in Figure 22.3 and 22.4. But the portion to the left of L_0 lies in Z_s. $\square$

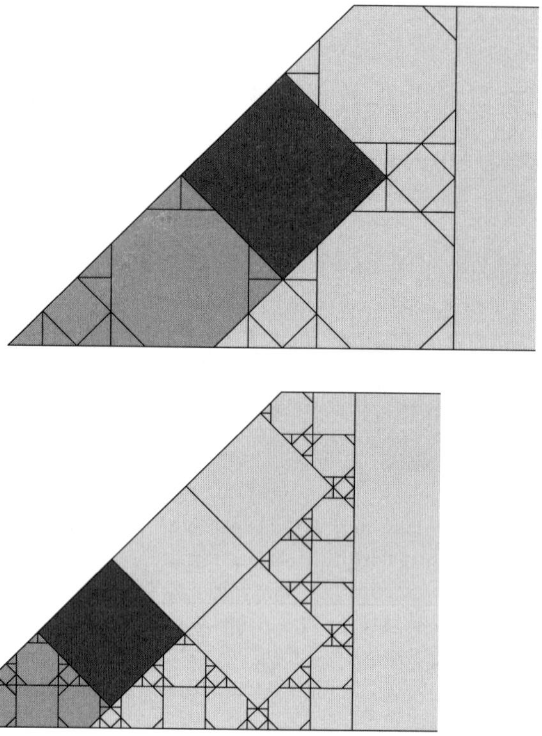

Figure 22.3: Bottom square (dark) and Z_s (light) for $s = 5/13, 16/39$.

22.4. Folding

Now we prove some results which will be useful when we prove Statement 3 of Theorem 1.7.

First we consider the case of symmetric pieces. Let K_s be a symmetric piece. We say that $S \subset \Lambda_s \sqcap K_s$ is a *tail* if S is compact and connected and intersects the base of K_s. We suppose that the tail has a chosen point on the base of K_s, and we call this point the *anchor*. In this case, we define $h(S)$ to be the maximum distance S rises away from the base.

LEMMA 22.5. *Suppose that K_s has a tail S with an anchor lying to the left of the centerline. Then K_s has another tail S' lying to the left of the centerline such that $h(S) = h(S')$.*

Proof: Let ρ be reflection in the centerline of K_s. Let S_L and S_R respectively denote the subsets of S lying to the left and to the right of the centerline. Then $S' = S_L \cup \rho(S_R)$ has all the desired properties. □

By symmetry, the same result holds with *right* replacing *left*. Our last result in this chapter is an amplification of Lemma 22.5. We say that an *anchored set* is a compact connected subset $S \subset \Lambda_s^0$ (the left half of Λ_s) having some point on the bottom edge of X_s. We define $h(S)$ to be the maximum distance S rises above the bottom edge of X_s. Let Z_s be the set from the Main Theorem.

LEMMA 22.6. *Suppose that $s < 1/2$ and $R(s) < 1/2$. Suppose S is an anchored set which does not cross the line D_s. Then there is another anchored set S' which either lies to the left of the leftmost zone centerline or else lies between between two successive zone centerlines. Moreover $h(S) = h(S')$.*

Proof: Choose an anchor point p of S. For ease of exposition, we consider the case when p lies strictly between two consecutive zone centerlines L and R. The portion of the tiling beneath D_s is a union of tiles in zones and square tiles in the pyramid. Therefore S lies in a union of zones. There is a partial ordering on the zones, according to the horizontal position of their centerlines. The line D_s is the right edge of the rightmost zone.

If S crosses the vertical line R then there is some zone ζ such that S crosses the centerline of ζ but S does not cross the right edge of ζ. Let $\rho : \zeta \to \zeta$ denote the reflection in the centerline of ζ. Let S_L (respectively S_R) denote the portion of S lying to the left (respectively right) of the centerline of ζ. If we replace S by

$$S_L \cup \rho(S_R) \tag{22.4}$$

then we retain the same h-value and the new anchored set does not cross the centerline of ζ. Repeating this trick finitely many times, we arrange that S does not cross R. Now we repeat the trick again, finitely many times, to arrange that S does not cross L. Figure 22.4 illustrates the construction. □

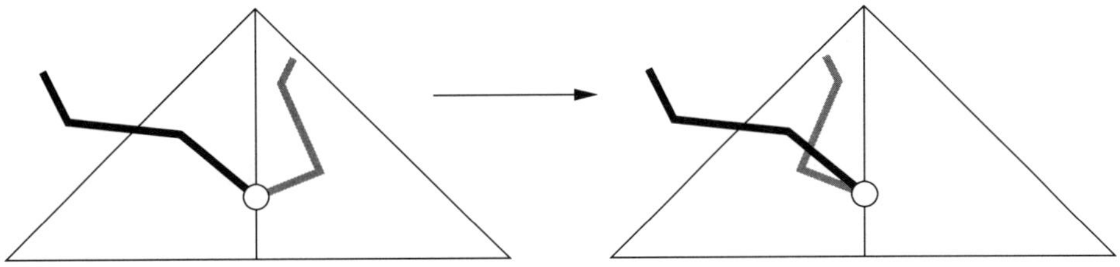

Figure 22.4: The folding trick.

LEMMA 22.7. *Suppose $s < 1/2$ and S is an anchored set which does not cross the line D_s. Then there exists an anchored subset $S' \subset Z_s$ such that $h(S) = h(S')$.*

Proof: When $t = R(s) > 1/2$, we proceed as in the proof of Lemma 22.4. The point here is that the map $R_H R_D R_H$ used in Equation 22.3 carries an anchored set to an anchored set. Hence, it suffices to consider the case when $t < 1/2$. Again, we can use the results about zones.

Starting with an anchored set S_1, we produce a new anchored set S_2 which either lies to the left of the leftmost vertical line in the octagrid or else lies between two consecutive vertical zone centerlines. Now we proceed as in the proof of Lemma 22.4. □

CHAPTER 23

The Forest Case

23.1. Reduction to the Loops Theorem

In this chapter we prove Statement 2 of Theorem 1.7, that Λ_s is a finite forest when s is irrational and $R^n(s) > 1/2$ only finitely often.

Let A_s and B_s be the symmetric pieces from §9. As in previous chapters, we let $\Lambda_s \sqcap A_s$ denote those points $p \in A_s$ such that every neighborhood of p contains infinitely many periodic tiles which belong to A_s. We define $\Lambda_s \sqcap B_s$ similarly.

LEMMA 23.1. *Both $\Lambda_s \sqcap A_s$ and $\Lambda_s \sqcap B_s$ are finite unions of arcs.*

Proof: By the Covering Lemma, $\Lambda_s \sqcap A$ is partitioned into finitely many tiles and finitely many ϵ-patches. If the patches are sufficiently small then the associated parameters are oddly even. But then the intersection of $\Lambda_s \sqcap A_s$ with each such patch is a finite union of arcs Hence $\Lambda_s \sqcap A_s$ is a finite union of arcs. The same goes for $\Lambda_s \sqcap B_s$. □

Letting Λ_s^0 denote the left half of Λ_s, we have

$$\Lambda_s^0 = (\Lambda_s^0 \sqcap A_s) \cup (\Lambda_s^0 \sqcap B_s). \tag{23.1}$$

Hence Λ_s is a finite union of arcs.

Now a finite union of arcs with no loops is a finite forest. Hence, the following result implies Statement 2 of Theorem 1.7.

THEOREM 23.2 (Loops). *Let $s \in (0,1)$ be any irrational number. Then Λ_s^0 contains no embedded loops.*

Remark: Statement 3 of Theorem 1.7 makes a much stronger statement when s is such that $R^n(s) > 1/2$ infinitely often. So, the Loops Theorem really only says something new when s is such that $R^n(s) > 1/2$ for finitely many n.

The rest of the chapter is devoted to proving the Loops Theorem.

23.2. Proof of the Loops Theorem

We proceed as in the proof of the Pinching Lemma. Say that a *counterexample* is a pair $\Omega = (\gamma, s)$ where γ is an embedded loop in Λ_s^0. Define

$$\lambda(\gamma) = \frac{\text{diam}(\gamma)}{\text{diam}(X_s^0)}. \tag{23.2}$$

Let M denote the supremum, taken over all values $\lambda(\gamma)$, where γ is a counterexample for some loop. We choose a counterexample (γ, s) and we explain why we get a contradiction if $\lambda(\gamma)$ is close to M.

Case 1: Suppose that $s < 1/2$. By Lemma 22.4, we can assume that $\gamma_s \subset \Lambda_s \sqcap Z_s^0$. But then $\gamma_t = \phi_s^{-1}(\gamma_s)$ is a loop in Λ_t with $\lambda(\gamma_t) \geq \heartsuit_s \lambda(\gamma_s)$, and we get the same contradiction as in the proof of the Pinching Lemma.

Case 2: Suppose $s > 1/2$. By the Inversion Lemma and the invariance of the quantity $\lambda(\gamma)$ under similarities, it suffices to take $s \in (1/2, \sqrt{2}/2]$. In this case, the reflection R_V maps the region to the right of V into Z_s^0 and the reflection R_D maps the region above D_s into Z_s^0. Figure 23.2 shows a fairly typical example.

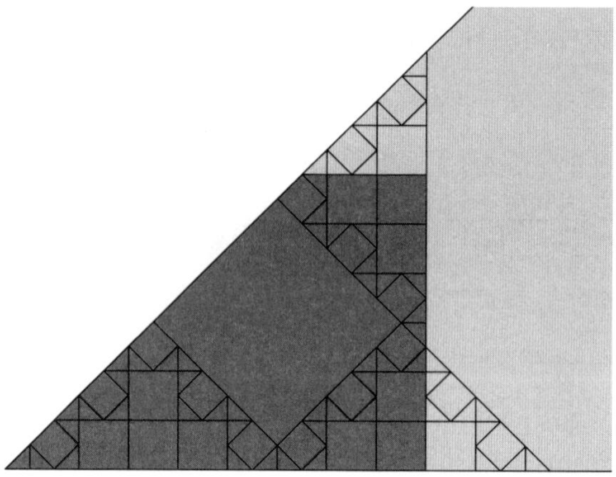

Figure 23.1: Z_s^0 (shaded) for $s = 11/17$.

By the Pinching Lemma, γ can intersect each of V and D_s at most once. Hence γ lies to one side or the other of each of these lines. So, by symmetry, we can assume that $\gamma \subset Z_s^0$. Now we proceed as in Case 1.

23.3. An Example

One might wonder about the combinatorics of the forests given by Statement 2 of Theorem 1.7. We did not investigate this in a systematic way, but we at least show that Λ_s can be a forest which is not just a union of arcs.

The left side of Figure 23.2 shows some of the unstable set Ξ_s for the parameter $s = 64/207$. The even expansion of s is

$$(3, 1, 2, 3, 1, 2, 2, 2, 2, 2).$$

Were we to consider the parameter with even expansion $(3, 1, 2, 3, 1, 2, 2, 2, \ldots)$, the limit set Λ_s would have the same combinatorial structure. The right side of Figure 23.2 shows the combinatorial structure of the portion of Ξ_s bounded by the lightly shaded tiles.

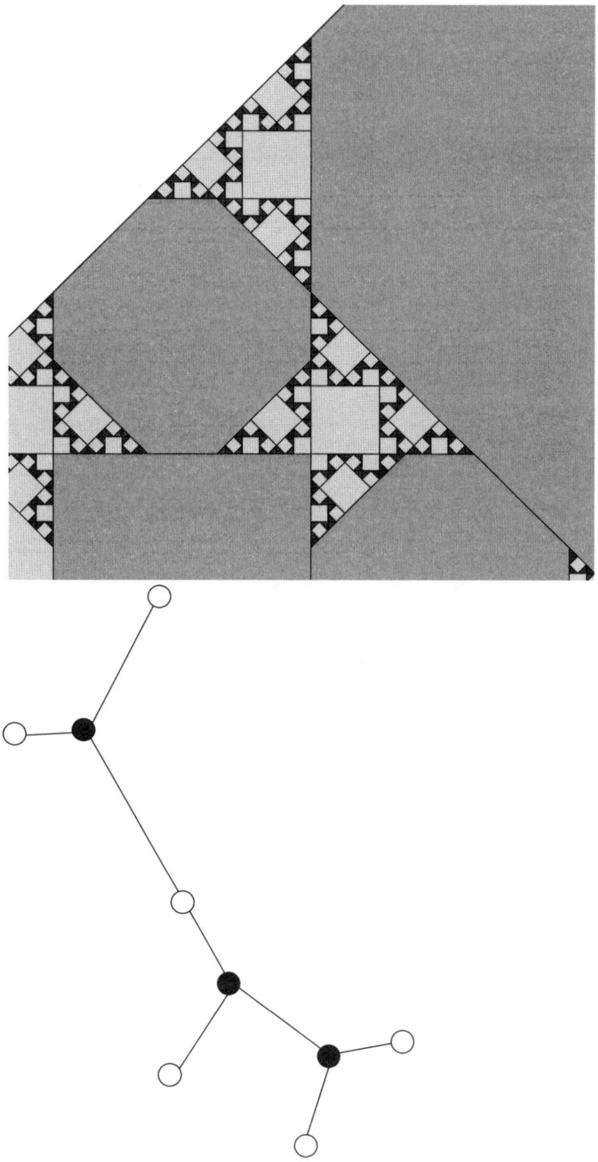

Figure 23.2: Ξ_s (dark) for $s = 64/207$.

The triple points tend to form at the vertices of the octagonal tiles. One might say that the octagons serve to splice the limit set together. We don't think that Λ_s can ever be a very complicated forest. As we discuss in §24.5 the octagonal tiles also chop the limit set into disconnected pieces and this prevents complicated trees from forming.

CHAPTER 24

The Cantor Set Case

We call $s \in (0,1)$ *octagonal* if $R^n(s) > 1/2$ infinitely often. In this chapter we prove Statement 3 of Theorem 1.7, that Λ_s is a Cantor set when s is octagonal. We mention again that §24.5 has an informal pictorial explanation of this result.

24.1. Unlikely Sets

Call $s \in (0,1)$ *octagonal* if $R^n(s) > 1/2$ infinitely often. We already know from §16.1 that To show that Λ_s is a Cantor set, it suffices to prove that Λ_s^0 is totally disconnected. We begin by ruling out a very unlikely kind of behavior for the set Λ_s.

Let C be a nontrivial connected subset of Λ_s^0. Let $\mathcal{K}$ be a patch cover, as in the Covering Lemma. What we mean is that $\mathcal{K}$ is a finite union of patches and tiles, as in the Covering Lemma. We call C *bad* with respect to $\mathcal{K}$ if C does not intersect the interiors of the images of any of the patches. That is, C is disjoint from the interiors of all the sets $\psi_j(K_j)$.

Call C *unlikely* if, for every $\epsilon > 0$, there is a patch covering $\mathcal{K}$ of scale less than ϵ, so that C is bad with respect to $\mathcal{K}$. So, C is bad with respect to an infinite sequence of patch covers, having scale tending to 0.

LEMMA 24.1. *Unlikely sets do not exist.*

Proof: Suppose $C \subset \Lambda_s^0$ is an unlikely component. Shrinking C if necessary, we take C to be one edge of the image of some patch. Lemma 15.2 tells us that each edge of each patch contains a tile edge. Hence, by Lemma 15.2, some segment of C lies in a tile boundary. Further shrinking C, we can assume that C is one edge of some tile τ_1.

The midpoint m belongs to Λ_s. Hence, there is some ϵ-patch (ψ, K, u) so that $m \in \psi(K)$ and τ_1 is disjoint from the interior of $\psi(K)$. One edge of $\psi(K)$ lies in the line containing C. Choosing ϵ small enough, we can assume that one edge of $\psi(K)$ is contained in C.

Some tile τ_2 of $\psi(K) \cap \Delta_s$ has an edge in C, by Lemma 15.2. But then some point of C is flanked on either side by the adjacent tiles τ_1 and τ_2 an hence cannot belong to Λ_s. This is a contradiction. □

24.2. Tails and Anchored Paths

We defined tails and anchored sets in §22.4.

LEMMA 24.2. *Suppose Λ_s^0 is not totally disconnected. Then, for all sufficiently large n, the parameter $u = R^n(s)$ is such that one of the symmetric pieces K_u has a tail.*

Proof: Let $\alpha = \alpha_s$ be a nontrivial connected component of Λ_s^0. By Lemma 24.1, once n is large enough and $u = R^n(s)$, we can find a patch (K, ψ, u) such that some point $p \in \alpha$ lies in the interior of $K^* = \psi(K)$ and some point q of α lies outside of K^*.

Now we "prune" α. Let U_n denote the $(1/n)$-tubular neighborhood of α. Since U_n is open and connected, there is some path $\beta_n \subset U_n$ which connects p to q. Evidently β_n intersects ∂K^*. Let α'_n denote the initial subpath of β_n which joins p to a point $r_n \in \partial K^*$. Passing to a subsequence, we can guarantee that $\{\alpha'_n\}$ converges in the Hausdorff topology to some subset α', and that r_n converges to some point $r \in \partial K^*$. By construction α' is a compact connected subset of both α and $\Lambda_s \sqcap K^*$ which joins p to r. The set $\psi^{-1}(\alpha')$ is a tail of K_u. $\square$

We call a compact connected subset $S \subset \Lambda_s^0$ *bottom* (respectively *left*) *anchored* if S contains a point on the bottom (respectively left) edge of X_s^0.

LEMMA 24.3. *Suppose Λ_s^0 is not totally disconnected. For all sufficiently large n, the parameter $r = R^n(s)$ is such that Λ_u^0 contains an anchored set.*

Proof: We choose n so large that K_u has a tail. Each symmetric piece K_u has the following property. Each edge of K_u either lies in ∂X_u^0, or else reflection in the centerline of K_u maps that edge into ∂X_u^0. So, once n is sufficiently large, and $u < 1/2$. The set Λ_u^0 contains a continuous path β having an endpoint $p \in X_u^0$.

If p does not lie in an edge of X_u^0 then p lies on the leftmost edge of the leftmost central tile of Δ_u. But then $p \in \partial P_u$, and some initial portion β' of β lies in P_u. But now the reflection R_D, the symmetry of $P_u \cap \Delta_u$, moves β so that it has an endpoint in $\partial X \cap \partial X_0$. We replace β by $R_D(\beta')$ if necessary.

If β is not yet anchored, it means that the endpoint p lies in the top edge of X_u^0. But then we repeat the same reflection trick, with A_u and R_H in place of P_u and R_D. The result $R_H(\beta')$ is bottom anchored. $\square$

24.3. Acute Crosscuts

Let K_s be a symmetric piece. Recall that $\Lambda_s \sqcap K_s$ is the set of points $p \in K_s$ such thta every neighborhood of p intersects infinitely many tiles of $\Delta_s \cap K_s$. We call $S \subset \Lambda_s \sqcap K_s$ an *acute crosscut* if S is compact and connected, and S contains points on two sides of K_s which make an acute angle with each other. The goal of this section is to prove the following result.

LEMMA 24.4. *If s is octagonal, then K_s cannot have an acute crosscut.*

If some symmetric piece K_s has an acute crosscut, then we can use bilateral symmetry to reduce to the case when Λ_s^0 has a connected subset C containing a point on the bottom edge of X_s and a point on the left edge of X_s. We abbreviate this by saying that the parameter s has an acute crosscut. We call the acute crosscut *small* if it is contained in Z_s and otherwise *big*.

LEMMA 24.5. *If $s < 1/2$ is octagonal, then s cannot have a big acute crosscut.*

Proof: Let $t = R(s)$. Let p be point of C contained in the left edge of X_s. If $p \in Z_s$ then C must contain points on opposite edges of Z_s. That means that some connected component of $\Lambda_s \sqcap Z$ contains points on opposite diagonal sides of Z_s.

But then a connected component of Λ_t contains points on opposite horizontal sides of X_t. When $t < 1/2$ we immediately contradict Corollary 20.5. When $t > 1/2$ we observe that the opposite horizontal sides of $X_t - O_t$ are separated by O_t. Here O_t is the central octagon.

The other case to consider is when $p \notin Z_s$. In this case $t < 1/2$ and C contains points on opposite sides $R_D\phi_s(A_t)$. But then there is some connected subset $C' \subset C$ such that $R_D(C') \subset \Lambda_s$ contains points on opposite sides of Z_s, as in the previous case, a contradiction □

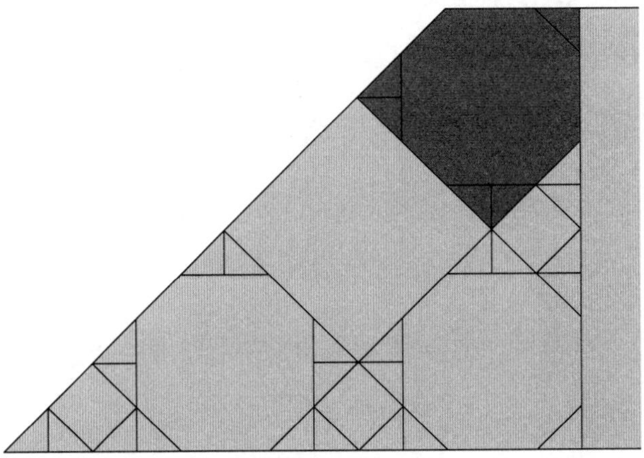

Figure 24.1: The set $R_D\phi_s(A_t)$ for $s = 5/13$ and $t = 3/10$.

LEMMA 24.6. *If $s \in (3/4, 1)$ is octagonal, then s does not have big acute crosscut.*

Proof: Let C be a crosscut which supposedly is not contained in Z_s. For s in this range, we make 3 observations.

- $Z_s \subset B_s$. Since $Z_s \subset B_s$, the symmetry R_V acts on Z_s. Hence, the set $Z_s^* = Z_s \cup R_V(Z_s)$ is well-defined and is contained in B_s.
- C must contain points on the tops and bottoms of Z_s^*. The point here is that any path from the bottom edge of X_s to the left edge of X_s must first rise above the line extending the top edge of Z_s^*.
- The tile $\phi_s(O_t)$ separates Z_s from $R_V(Z_s)$ in the sense that any path connecting these two sets must cross the line extending the top of the tile.

Figure 24.1 illustrates this for a representative parameter.

All these observations imply that there is some smaller connected subset

$$C' \subset C' \subset Z_s^*$$

having points on both the top and the bottom of $Z - s^*$. Applying R_V if necessary, we can assume that $C' \subset Z_s$. But then $\phi_t(C')$ connects opposite horizontal sides of X_t and lies in Λ_t. This contradicts Corollary 20.5. □

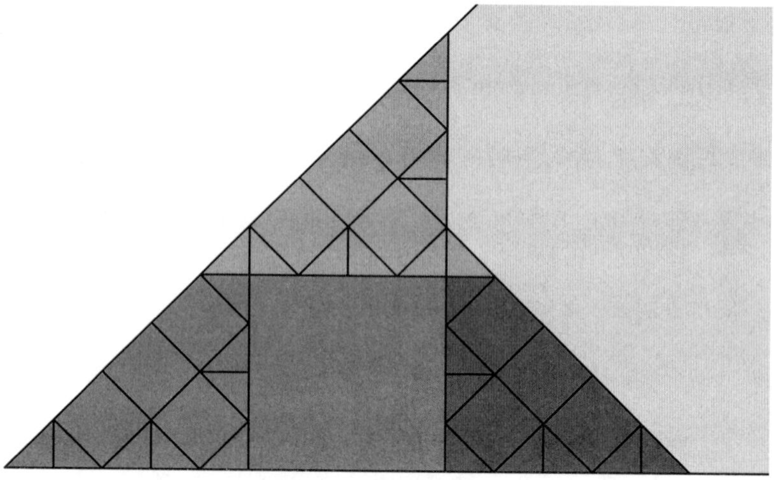

Figure 24.2: Z_s^* (medium/dark) and B_s (shaded) for $s = 7/9$.

LEMMA 24.7. *If $s \in (2/3, 3/4)$ is octagonal, then s does not have a big acute crosscut.*

Proof: For parameters in this range, the tile L_t to the left of the central tile in Δ_t is an octagon and $\phi_s(L_t)$ separates X_s^0 into three regions. This is the darkly shaded octagon in Figure 24.3.

If our crosscut Z_s does not lie in Z_s, is must contain the point

$$p = O_s \cap \phi_s(L_t). \tag{24.1}$$

Here O_s is the central tile of Δ_s. Hence p lies in a nontrivial connected component of $\Lambda_s \sqcap Q_s$. But then $R_E(p)$ lies in a nontrivial connected component of $\Lambda_s \sqcap Q_s$. But then $R_D R_E(p)$ lies in a nontrivial connected component of Λ_s. But this last point is the bottom left vertex of X_s. The same argument as in Lemma 21.1, in the octagonal case, contradicts this: There are infinitely many octagonal tiles wedged into this corner. □

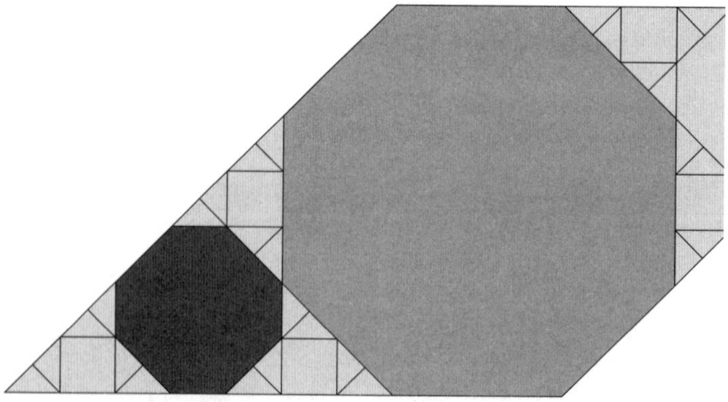

Figure 24.3: $\phi_s(L_t)$ (dark) and O_s (light) for $s = 7/10$ and $t = 3/10$.

LEMMA 24.8. *If $s \in (1/2, 2/3)$ then s does not have a big acute crosscut.*

Proof: Let $\sigma = 1/(2s) \in (3/4, 1)$. Let ρ be the similarity from the Inversion Lemma, so that $\rho(X_\sigma) = X_s$. We know that σ does not have any big acute crosscuts. Therefore, an acute crosscut for s is contained in $\rho(Z_\sigma)$. But $\rho(Z_\sigma \subset$ for $s \in (1/2, 2/3)$. Figures 24.4 and 24.5 show some examples.

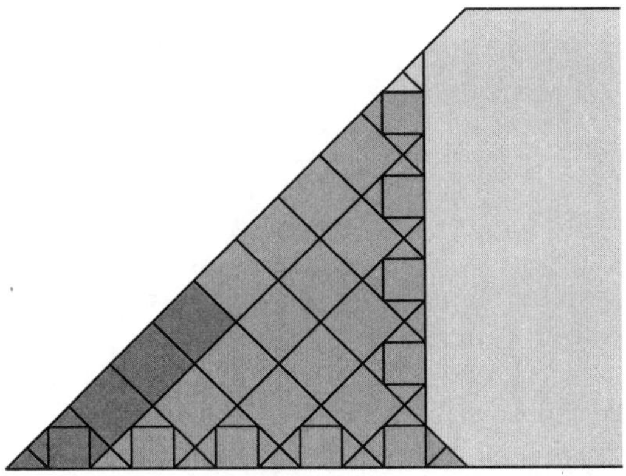

Figure 24.4: $\rho(Z_\sigma)$ (dark) and Z_s (shaded for $s = 11/20$.

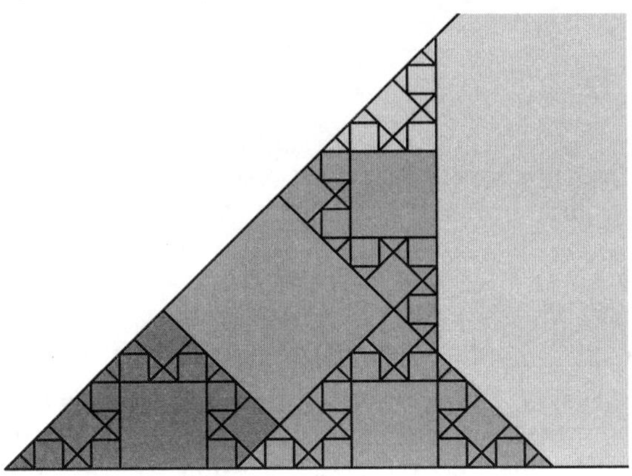

Figure 24.5: $\rho(Z_\sigma)$ (dark) and Z_s for $s = 19/30$.

The following lemma contradicts the previous ones, unless no octagonal parameter has an acute crosscut.

LEMMA 24.9. *If an octagonal parameter s has a crosscut, then some other octagonal parameter has a big acute crosscut.*

Proof: If s has an acute crosscut contained in Z_s and $t = R(s)$ also has an acute crosscut. If the acute crosscut for t lies in Z_t we can repeat the procedure. Every one or two steps of the procedure, the distance between the endpoints of the crosscut increases by a factor of at least $\sqrt{2}$. So, eventually we reach a stage where the crosscut is not contained in Z. □

24.4. The Main Argument

Suppose that there is some octagonal parameter s such that Λ_s is not totally disconnected. Then we can find some new octagonal parameter $u < 1/2$ where Λ_u contains an anchored set. For any bottom anchored set α, we define $h(\alpha)$ to be the maximum distance α rises above the bottom edge of X_s. This is exactly as in the previous chapter. When α is left anchored, we define $h(\alpha)$ to be the maximum distance α moves away from the left edge of X_s. Similar to what we did in previous chapters, we define

$$\lambda(\alpha) = \frac{h(\alpha)}{\text{diam}(X_s^0)}. \tag{24.2}$$

We let M denote the supremum value taken over $\lambda(\alpha)$, where α ranges over all bottom or bottom anchored sets α in Λ_s, for $s \in (0,1)$.

For each anchored set, we choose some point in the relevant edge of X_s. If there are several choices, we choose arbitrarily. We call this point *the anchor point*. We divide the argument into 4 cases which exhaust the possibilities.

Let D_s, E_s, H, V denote the lines of symmetry from §9.

Case 1. Suppose that

- $s < 1/2$.
- α_s is bottom anchored.
- α_s lies beneath D_s.

By Lemma 22.7, we can assume that $\alpha_s \subset Z_s^0$. Pruning α_s as in the proof of Lemma 24.2, we can assume that $\alpha_s \subset \Lambda_s \sqcap Z_s^0$. But then $\alpha_t = \phi_s^{-1}(\alpha_s)$ is a left anchored set in Λ_t, and $\lambda(\alpha_t) = \heartsuit_s \lambda(\alpha_s)$. Here we get the same contradiction as in the proofs of the Pinching Lemma and the Loops Theorem.

Case 2. Suppose that s satisfies the same conditions as in Case 1, except that S crosses D_s. Since B_s lies entirely beneath D_s, and the left edge of B_s lies in ∂X_s, we see that α_s contains a point on the right edge of B_s. Pruning α_s as in the proof of Lemma 24.2, we produce an acute crosscut for B_s. This contradicts Lemma 24.4. Hence the anchor of α_s lies in A_s.

By Corollary 20.5, we have $\alpha_s \subset A_s$. For s in the range here, we have $A_s \subset P_s$. Since R_H is the symmetry of $A_s \cap \Delta_s$ and R_D is the symmetry of $P_s \cap \Delta_s$, the containments above imply that $\beta_s = R_D R_H R_D(\beta_s) \subset \Lambda_s$ is a bottom anchored set such that $h(\beta_s) = h(\alpha_s)$. By construction, the anchor point of β_s lies in B_s. But the contradicts the preceding paragraph.

Case 3. Suppose that

- $s < 1/2$.
- α_s is left anchored.

When $t = R(s) > 1/2$, the octagonal tile $\phi_s(O_t)$, shown in Figure 5.1, forces a left anchored set to lie entirely in Z_s. Figure 24.6 shows a typical picture. Now we proceed as in Case 1.

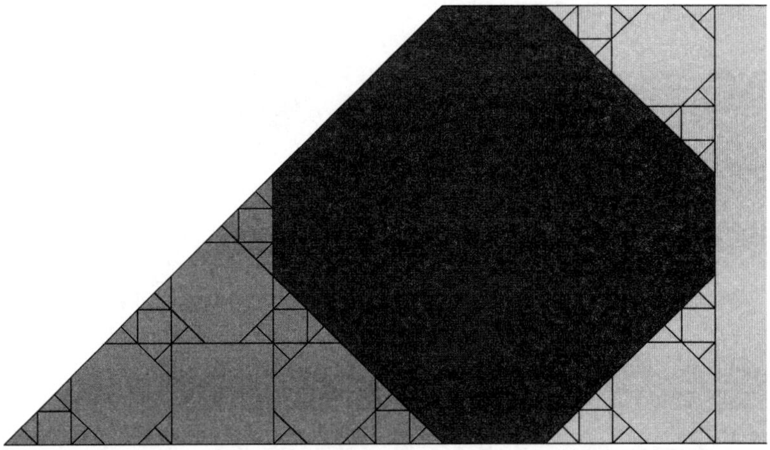

Figure 24.6: Z_s (light) and $\phi_s(O_t)$ (dark) for $s = 13/42$.

On the other hand, suppose $t < 1/2$. We claim that $\alpha_s \subset Z_s^0$. Otherwise, α_s contains points on the opposite diagonal edges of Z_s^0. Pruning α_s as in the proof of Lemma 24.2, we see that there is a left-anchored subset $\alpha_s' \subset \Lambda_s \sqcap Z_s^0$ which contains points on the two diagonal edges of Z_s^0. But then $\alpha_s = \phi_s^{-1}(\alpha_s)$ is a compact connected set in Λ_t which contains points on both the top and the bottom edges of X_t^0. This contradicts Corollary 20.5.

Case 4. Suppose that $s > 1/2$. Using the Inversion Lemma, it suffices to consider the case when α_s is bottom anchored. Note that the bottom edge of X_s^0 is contained in B_s. So, the anchor point of α_s lies in B_s. If α_s exits B_s, we contradict Lemma 24.4, as in Case 2. Hence, α_s does not exit B_s.

By Lemma 22.5, we can assume that α_s lies to the left of V, the vertical line of symmetry of B_s, without changing $h(\alpha_s)$. But then α_s is anchored in Z_s^0 and α_s can only exit Z_s^0 through the top edge. If $\alpha_s \subset Z_s^0$, then we proceed as in Case 1, and we produce a new anchored set α_t with $\lambda(\alpha_t) > \lambda(\alpha_s)$. This reduces to Cases 1-3.

If α_s exits Z_s^0 through the top edge, then we prune α_s as in the proof of Lemma 24.2, and find an anchored set $\alpha_s' \subset \Lambda_s \sqcap Z_s^0$ which contains points on both horizontal edges of Z_s^0. But then $\phi_s^{-1}(\alpha_s') \subset \Lambda_t$ is a compact connected subset of Λ_t which contains points on both horizontal edges of X_t^0. This contradicts Corollary 20.5.

24.5. Pictorial Explanation

Our proof above somewhat clouds the intuitive reason why Λ_s is a Cantor set when s is an octagonal parameter. Here we give an informal and intuitive pictorial explanation.

We show two pictures for the parameter $s = 401/1092$. (This is kind of a rational approximation to an octagonal parameter.) The parameter s has even expansion

$$(2, 2, 2, 3, 1, 2, 2, 2, 3, 1, 4).$$

In the rational setting we work with Ξ_s, the union of triangular tiles of Δ_s.

As predicted by Theorem 12.1, there are octagonal tiles of two different sizes. The larger octagonal tiles are lightly shaded in Figure 24.7.

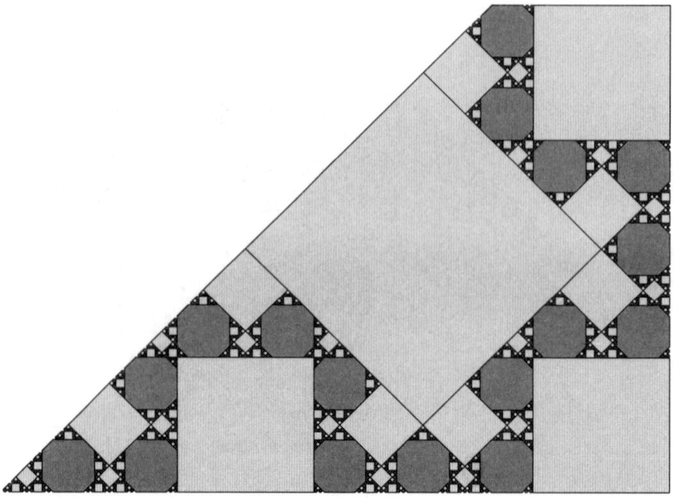

Figure 24.7: Δ_s^0 for $s = 401/1092$.

These octagonal tiles attach on one side to a union of boxes, and on the other to a union of diamonds. The union of boxes, diamonds, and large octagons chops up X_s^0 into a large number of small regions which are disconnected from each other. Ξ_s^0 must lie in the union of these regions. This forces the connected components of Ξ_s^0 to be pretty small.

Figure 24.8 shows a closeup of one of the small regions. This time, the small dark octagonal tiles run through the small region, attaching themselves to diamonds and big octagons on one side and boxes on the other. The union of these tiles chops up each small region into a number of tiny regions, each disconnected from each. Ξ_s^0 is forced to lie in these tiny regions.

Figure 24.8: Δ_s^0 for $s = 401/1092$.

Were s an octagonal parameter, this kind of pattern would continue forever, and in a recursive way, completely chopping up Λ_s into a Cantor set.

CHAPTER 25

Dynamics in the Arc Case

25.1. The Main Result

Let $s \in (0,1)$ be an oddly even parameter. We will take s to be irrational, but the reader should bear in mind that we will also consider the rational case below.

Let $(k_0, k_1, k_2, \ldots)$ be the even expansion of s. By the Insertion Lemma, the value of the first digit $k_0 \in \{2, 4, 6, \ldots\}$ has no influence on the system. When we take $k_0 = 2$ we have $s \in [1/2, 1/3)$. We define

$$a_s = \frac{1}{k_1 + \frac{1}{k_2 \ldots}}, \qquad b_s = 1 - 2a_s. \tag{25.1}$$

That is, the continued fraction of a_s is $(0, k_1, k_2, \ldots)$. The reason that $a_s < 1/2$ is that $k_1 > 1$.

The set $\mathcal{E}$ of oddly even parameters has the form $C - C'$, where C is a Cantor set and C' is a countable set. The maps $a \to a_s$ and $s \to b_s$ are continuous functions on $\mathcal{E}$, and they both map $\mathcal{E}$ into $\boldsymbol{R} - \boldsymbol{Q}$.

Remark: When the continued fraction of s is $[0, c_1, c_2, c_3, \ldots]$ the even expansion of s satisfies $k_j = c_j$ for j odd and $k_j = 2c_j$ when j is even.

Figure 25.1 shows a 10-interval IET, defined in a union of two intervals, each having length 2. The top half shows the first partition and the second half shows the bottom partition. The top labellings of each half label the intervals in a combinatorial sense and the bottom labellings of each half give the lengths of the intervals. In the figure, we have set $a = a_s$ and $b = b_s$. We call this IET I_s.

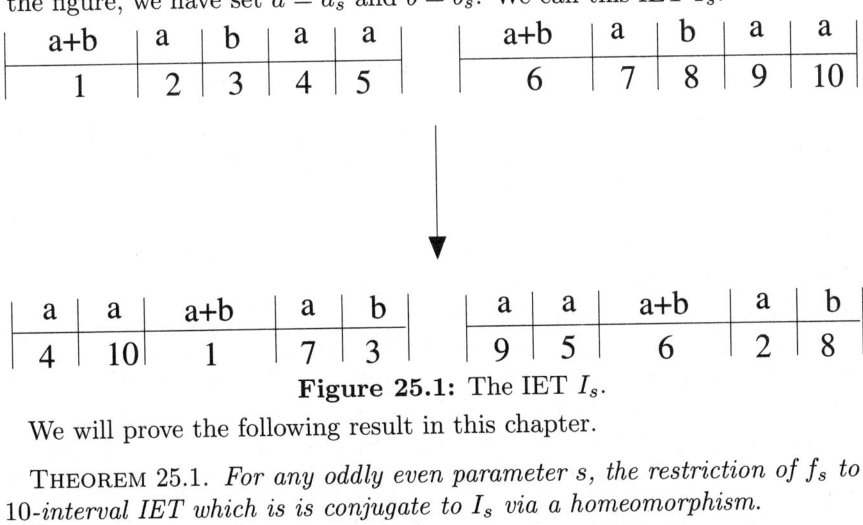

Figure 25.1: The IET I_s.

We will prove the following result in this chapter.

THEOREM 25.1. *For any oddly even parameter s, the restriction of f_s to Λ_s is a 10-interval IET which is is conjugate to I_s via a homeomorphism.*

The IET in Figure 25.1 commutes with swapping the two intervals. Thus, there is a quotient IET, $\overline{I}_s$ defined on an interval of length 2. Figure 25.2 shows the picture.

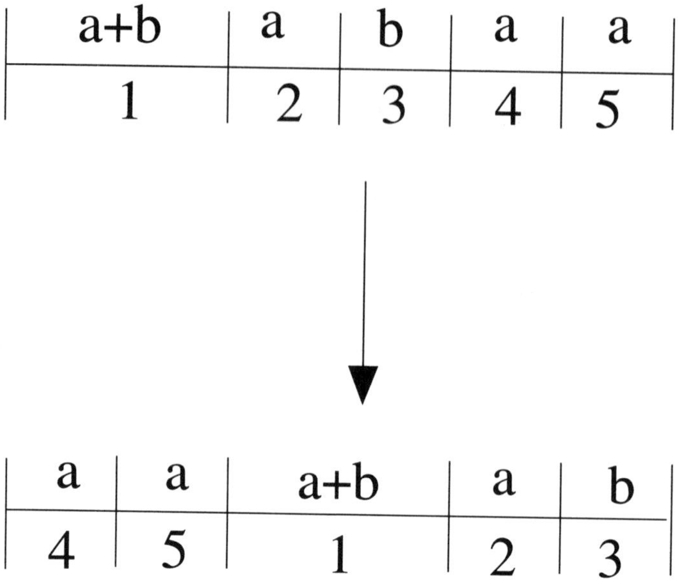

Figure 25.2: The IET $\overline{I}_s$.

The map $\overline{I}_s$ is just an irrational rotation with rotation number a_s. Thus, I_s is a $\boldsymbol{Z}/2$ extension of the rotation $\overline{I}_s$ which has rotation number a_s. Taking into account our remark above, we see that, in terms of continued fraction expansions,

$$s = [0, c_1, c_2, c_3, ...] \implies a_s = [0, 2c_2, c_3, 2c_4, c_5, 2c_6, c_7, ...]. \tag{25.2}$$

Thus, Theorem 1.8 is an immediate corollary of Theorem 25.1.

The rest of the chapter is devoted to proving Theorem 25.1. We will first establish the combinatorics of the restriction $f_s|\Lambda_s$, and then we will use counting and rational approximation ideas to pin down the geometry.

25.2. Intersection with the Partitions

For each parameter s, the parallelogram X_s has two partitions $\mathcal{P}_s$ and $\mathcal{P}_s^*$ which define the PET $f_s : X_s \to X_s$. When s is oddly even, and the first term in the even expansion is 2, we have $s \in (1/3, 1/2)$.

In this section we show that Λ_s intersects each piece of $\mathcal{P}_s$ in an arc. The reader should bear in mind that the same proof works, in therr rational case, for the union L_s of long sides of the unstable triangles.

Since f_s preserves Λ_s and maps the pieces of $\mathcal{P}_s$ to $\mathcal{P}_s^*$, we get the same result for $\mathcal{P}_s^*$. Since everything in sight commutes with reflection in the origin, it suffices to prove our result for the 5 pieces $P_s^1, ..., P_s^5$ of $\mathcal{P}_s$ which lie in the left half of X_s. The pieces in the picture are ordered according to the order they are visited by L_s, starting from the lower left corner. Thus P_1 is the light trapezoid, P_2 is the small dark triangle, P_3 is the medium pentagon, P_4 is the light triangle, and P_5 is the dark pentagon.

Figure 25.3: The left half of the partition for $s = 7/17$.

We consider the arc Λ_s^0, the portion of Λ_s lying to the left of the origin. One endpoint of Λ_s^0 is the bottom left vertex of X_s, and the other endpoint if the top left vertex of X_s. We orient Λ_s^0 from the bottom to the top.

We say that Λ_s^0 *crosses* a polygon Q if there are 3 points $p_1, p_2, p_3 \in \Lambda_s^0$, coming in the order of the parameterization, such that $p_1, p_3 \in Q$ and $p_2 \notin Q$. In other words, Λ_s^0 exits Q and then re-enters Q at a later time. If Λ_s^0 does not cross Q, then $\Lambda_s^0 \cap Q$ is an arc.

LEMMA 25.2. *Λ_s^0 crosses neither $P_s^1 \cup P_s^2$ nor $P_s^3 \cup P_s^4 \cup P_s^5$.*

Proof: We check by direct inspection of the polyhedra involved, and listed in Part IV, that $A_s = P_s^1 \cup P_s^2$ and $B_s = P_s^3 \cup P_s^4 \cup P_s^5$. Figure 25.2 shows the picture for a typical parameter. It follows from the Shield Lemma that Λ_s^0 does not cross A_s. Note that $\partial B_s - \partial X_s \subset \partial A_s$, and the final point of Λ_s^0 lies in A_s. Hence, if Λ_s^0 crosses B_s then Λ_s^0 also crosses A_s. □

LEMMA 25.3. *Λ_s^0 crosses neither P_s^1 nor P_s^2.*

Proof: It suffices to prove that Λ_s^0 does not cross P_s^2. Referring to the Main Theorem, we have $P_s^2 = R_V \circ \phi_s(B_t)$. Here $t = R(s)$. If Λ_s^0 crosses P_s^2 then Λ_s^0 also crosses $\phi_s(B_t)$. But then the arc Λ_t^0 crosses B_t. This does not happen, by the arguments given in the proof of Lemma 25.2. Since Λ_t^0 does not cross B_t, we see that Λ_s^0 does not cross P_s^2. □

LEMMA 25.4. *Λ_s^0 does not cross P_s^5.*

Proof: Referring to the Main Theorem, and to §9, we have

$$P_s^5 = Q \cup R_D \circ \phi_s(A_t),$$

where R_D is reflection in the diagonal fundamental line of symmetry and Q is a finite union of square periodic tiles. If Λ_s^0 crosses P_s^5 then Λ_s^0 also crosses $\phi_s(A_t)$. But then Λ_t^0 crosses A_t. This does not happen. □

LEMMA 25.5. *Λ_s^0 crosses neither P_s^3 nor P_s^4.*

Proof: We compute $P_s^4 = R_D R_H(P_s^2)$. Since Λ_s does not cross P_s^2, it does not cross P_s^4 either. Since Λ_s does not cross 4 out of the 5 pieces, it does not cross the 5th piece either. □

Now we know that Λ_s^0 also intersects the each of $P_s^6, ..., P_s^{10}$ in an arc. The upshot of the discussion is that the restriction $f_s|\Lambda_s$ is a 10-interval IET for all oddly even parameters s, and the combinatorics of the IET does not depend on the parameter. The combinatorics can be gleaned by looking at a single parameter, and one gets the combinatorics shown in Figure 25.1.

25.3. The Rational Case

The results from the previous section hold when s is rational and L_s replaces Λ_s. We first prove 25.1 in the rational case. Let s be an oddly even rational parameter. Euclidean arc length gives a measure μ_s on L_s. We normalize so that $\mu_s(L_s) = 4$. Let A_s and B_s be the symmetric pieces from §9.

LEMMA 25.6. *$\mu_s(A_s \cap L_s) = \mu_s(B_s \cap L_s) = 1$.*

Proof: This is one case of Lemma 18.3. □

The pieces $P_s^1, ..., P_s^{10}$ refer to the partition for f_s.

LEMMA 25.7. *$\mu_s(P_s^2 \cap L_s) = \mu_s(P_s^4 \cap L_s) = \mu(P_s^5 \cap L_s)$.*

Proof: Referring to the reflections from §9 we have

$$R_H \circ R_D(P_s^2 \cap L_s) = P_s^4 \cap L_s, \tag{25.3}$$

and acts in such a way as to preserve μ_s. This proves the first of the identities,
For the second identity, we use the Main Theorem. We have

$$\mu_s(L_s \cap P_s^5) = \lambda\mu_t(L_t \cap A_t), \tag{25.4}$$

and

$$\mu_s(L_s \cap P_s^2) = \lambda\mu_t(L_t \cap B_t). \tag{25.5}$$

Here λ depends on s but is the same in both equations. Our result now follows from Equations 25.4 and 25.5, and the previous lemma. □

At this point, we set

$$a_s = \mu_s(P_s^2 \cap L_s), \qquad b_s = \mu_s(P_s^3 \cap L_s). \tag{25.6}$$

LEMMA 25.8. *$\mu_s(P_s^1 \cap L_s) = a_s + b_s$.*

Proof: We also know that

$$A_s = P_s^1 \cup P_s^2, \qquad B_s = P_s^3 + P_s^4 + P_s^5.$$

Combining the results we have already proved, we get

$$\mu_s(P_s^1 \cap L_s) + a_s = \mu_s(A_s \cap L_s) = \mu(B_s \cap L_s) = a_s + a_s + b_s.$$

To complete the proof, we solve for the first term. □

LEMMA 25.9. $b_s = 1 - 2a_s$

Proof: We have $4 = \mu_s(L_s) = 4a_s + 2b_s$. □

All in all, we get the labels in Figure 25.1. It only remains to work out a_s as a function of s. Let $\widehat{a}_s$ be the number of unstable tiles in $P_s^2 \cap \Delta_s$. Likewise define $\widehat{b}_s$. Let

$$\widehat{c}_s = 2\widehat{a}_s + \widehat{b}_s. \tag{25.7}$$

Let $t = R(s)$ and $u = R(t)$. (When $u = 0$ we set to zero all quantities associated to u.)

LEMMA 25.10. $\widehat{a}_s = \widehat{c}_t$.

Proof: It follows from symmetry and the Main Theorem that

$$R_D(\Delta_s \cap P_s^5) = \phi_s(\Delta_t \cap A_t).$$

The number of unstable tiles in the set on the left is $\widehat{a}_s$ and the number of unstable tiles on the right is the number of unstable tiles is $\widehat{c}_t$. □

Suppose now that s has even expansion $2, k_1, k_2, \ldots$. Rule 1 from §18 gives us

$$\widehat{c}_s = k_1\widehat{c}_t + \widehat{c}_u. \tag{25.8}$$

Since $2a_s + b_s = 1$, we have

$$a_s = \frac{\widehat{a}_s}{\widehat{c}_s} \tag{25.9}$$

This gives us the recurrence relation

$$a_s = \frac{\widehat{c}_t}{\widehat{c}_s} = \frac{\widehat{c}_t}{k_1\widehat{c}_s + \widehat{c}_u} = \frac{1}{k_1 + (\widehat{c}_u/\widehat{c}_t)} = \frac{1}{k_1 + a_t}.$$

In short

$$a_s = \frac{1}{k_1 + a_t}. \tag{25.10}$$

We check the formula for Theorem 25.1 when $s = 1/2$. The result then follows for $s = 1/2n$ by the Insertion Lemma. The general rational case now follows from induction on the length of the R-orbit of s and Equation 25.10.

This completes the proof of Theorem 25.1 in the rational case.

25.4. Measures of Symmetric Pieces

Before we turn to the proof of Theorem 25.1 in the irrational case, we we record the formulas

- $\mu(A_s \cap L_s) = 2a_s + b_s$.
- $\mu(B_s \cap L_s) = 2a_s + b_s$.
- $\mu(P_s \cap L_s) = 3a_s + 2b_s$.
- $\mu(Q_s \cap L_s) = a_s$.

The formula for Q_s comes from the fact that R_V carries Q_s to the piece P_s^2 in our partition. The formula for P_s then follows from the fact that $P_s \cup Q_s = A_s \cup B_s$.

For any two symmetric pieces K_s and K'_s,

$$\mu(K_s \cap L_s) \geq \lambda_s \mu(K'_s \cap L_s), \qquad \lambda_s = \frac{a_s}{3a_s + 2b_s}. \tag{25.11}$$

In other words, the measure of one symmetric piece controls the measure of them all.

LEMMA 25.11. *Let $u = R^k(s)$. Let N_u denote the total number of patches in the patch covering of L_s^0 associated to the parameter u. Then, for any patch image K', $\mu_s(K' \cap L_s) \in [\epsilon_1, \epsilon_2]$, where*

$$\epsilon_1(k, s) = \frac{2\lambda_u}{N_u}, \qquad \epsilon_2(k, s) = \frac{2}{\lambda_u N_u}.$$

Proof: This is an immediate consequence of Equation 25.11 and the definition of a patch. The point is that μ_s assigns a value to each patch image which is at least λ_u, and at most λ_u^{-1} of the value it assigns to any other patch. □

The important point for is as follows. If $\{s_n\}$ is a sequence of rational parameters converging to s, then the sequence $\{\epsilon_j(k, s_n)\}$ converges to $\epsilon_j(k, n)$. In particular, the quantities in the sequence are uniformly bounded away from 0 and uniformly bounded away from ∞.

25.5. Controlling the Measures

It only remains to prove Theorem 25.1 in the irrational case. Let $\{s_n\}$ be a sequence of oddly even rational parameters converging to s. Let $\Lambda = \Lambda_s$ and $L_n = L_{s_n}$. Likewise define f and f_n. Let $\mu_n = \mu_{s_n}$ be the measure considered above. The following lemma provides the control we need in order to take a limit of these measures.

LEMMA 25.12. *Let $\{J_n\}$ be a sequence of arcs such that $J_n \subset L_n$. Then $\mu_n(J_n) \to 0$ if and only if* $\operatorname{diam}(J_n) \to 0$.

Proof: Suppose that the diameter of J_n does not tend to 0. Trimming J_n if necessary, we can assume that there is a patch (K_n, ψ_n, u_n) such that

$$J_n = \Lambda_n \cap \psi_n(K_n), \qquad u_n = R^{k_n}(s_n). \tag{25.12}$$

The lower bound on the diameter of J_n forces an upper bound on the sequence $\{k_n\}$. So, we can pass to a subsequence so that $k_n = k$ for all n. But

$$\mu_n(J_n) \geq \epsilon_1(k, s_n). \tag{25.13}$$

As we remarked at the end of the last section, the quantity on the right is uniformly bounded away from 0.

On the other hand, suppose that $\operatorname{diam}(J_n) \to 0$. Given the shapes of the symmetric pieces, they cannot pack too closely together. If we fix k, then J_n can only be contained in, say, 8 patch images of the patch cover associated to $R^k(s_n)$ once n is sufficiently large. But then

$$\mu_n(J_n) \leq 8\epsilon_2(k, s_n), \qquad n > n_k. \tag{25.14}$$

Hence

$$\limsup_{n\to\infty} \mu_n(J_n) \le 8\epsilon_2(k,s). \tag{25.15}$$

But k is arbitrary. Letting $k \to \infty$ we see that $\mu_n(J_n) \to 0$. □

Actually, we will not take the limit of the measures (though we could). Rather, we will work with the homeomorphisms defined by these measures.

25.6. The End of the Proof

We already know Theorem 25.1 in the rational case. Let $h_n : L_n \to I_1 \cup I_2$ be the homeomorphism obtained by integrating the measure μ_n. By the rational case of Theorem 25.1, we have $\{h_n f_n h_n^{-1}\} = I_n$. Given the fact that the continued fractions describing the lengths converge, we see that $h_n f_n h_n^{-1} \to I$. To finish the proof of Theorem 25.1, we just have to prove that $\{h_n\}$ converges uniformly to a homeomorphism $h : \Lambda \to I_1 \cup I_2$ on a subsequence. Given this convergence, we have $hfh^{-1} = I$, as desired.

Pick some point $x_0 \in \Lambda$. On a subsequence $\{h_n(x_0)\}$ converges. We choose some subsequential limit and define this to be $h(x_0)$. Suppose we make the same definition for another point x_1.

LEMMA 25.13. $h(x_0) \neq h(x_1)$.

Proof: Let J_n be an arc of L_n chosen so the endpoints of J_n converge to x_0 and x_1. The condition $h(x_0) = h(x_1)$ implies that $\mu(J_n) \to 0$. This contradicts Lemma 25.12. □

LEMMA 25.14. *Given any $\epsilon > 0$ there is some $\delta > 0$ so that $\|x_0 - x_1\| < \delta$ implies that $|h(x_0) - h(x_1)| < \epsilon$.*

Proof: If this result is false, then we can find a sequence $\{J_n\}$ of arcs, as in the previous result, such that $\text{diam}(J_n) \to 0$ but $\mu_n(J_n) > \epsilon$ for all n. This contradicts Lemma 25.12. □

Now, we choose a dense sequence $\{x_m\}$ of Λ and pass to a subsequence so that $\{h_n(x_m)\}$ converges for all m. We define $h(x_m) = \lim_{n\to\infty} h_n(x_m)$. The lemmas above imply that h maps this dense set injectively to another dense set. Moreover, h preserves the ordering of these points. Hence h extends to a homeomorphism from Λ to $I_1 \cup I_2$, and $h_n \to h$ uniformly. But this is all we need to finish the proof of Theorem 25.1 in the irrational case.

Part 5

Computational Details

Here is an overview of this part of the monograph:

- In §26 we describe the general computational methods we use for our calculations. The basic idea is to reduce everything to integer linear algebra calculations in $\boldsymbol{R}^3$.
- In §27 we describe Calculations 1-12 in detail. Calculations 1-8 can be launched from our program OctaPET and Calculations 9-12 can be launched from our program BonePET.
- In §28 we list the raw data used in our calculations. All the data is also listed in the files for our programs, OctaPET and BonePET. At the beginning of §28 we describe where the data resides.

CHAPTER 26

Computational Methods

26.1. The Fiber Bundle Picture

Our strategy is to reduce all our calculatons to statements about convex lattice polyhedra. In this section, we explain the main idea behind this reduction, for Calculations 1-8. We take the parameter in the interval $[1/4,2]$, though sometimes we will take subintervals of $[1/4,2]$.

In our standard normalization, X_s be the parallelogram with vertices

$$(\epsilon_1+\epsilon_2 s, \epsilon_2 s), \qquad \epsilon_1,\epsilon_2 \in \{-1,1\}. \tag{26.1}$$

We define

$$\mathcal{X} = \{(x,y,s)|\ (x,y) \in X_s)\} \subset \boldsymbol{R}^2 \times [1/4,2]. \tag{26.2}$$

The space $\mathcal{X}$ is both a convex lattice polyhedron and a fiber bundle over $[1/4,2]$ such that the fiber above s is the parallelogram X_s. See §28.2 for the list of vertices of $\mathcal{X}$. The maps $f_s : X_s \to X_s$ piece together to give a fiber-preserving map $F:\mathcal{X}\to\mathcal{X}$.

LEMMA 26.1. *The map F is a piecewise affine map.*

Proof: Consider some point $p\in X_s$. There is some vector

$$V_s = (A+Bs, C+Ds)$$

Here A,B,C,D are integers such that $f_s(p)=p+V_s$. If we perturb both V and s, the integers A,B,C,D do not change. So, in a neighborhood of p in $\mathcal{X}$, the map F has the form

$$(F(x,y,s) = (x+Bs, y+Ds, s)+(A,C,0).$$

This is a locally affine map of $\boldsymbol{R}^3$. $\square$

Let $\mathcal{X}(S)$ be the subset of our fiber bundle lying over a set $S\subset[1/4,2]$. Given the nature of the map, we find it useful to split our fiber bundle into 3 pieces, namely

$$\mathcal{X} = \mathcal{X}[1/4,1/2] \cup \mathcal{X}[1/2,1] \cup \mathcal{X}[1,2]. \tag{26.3}$$

A *maximal domain* of $\mathcal{X}(S)$ is a maximal subset on which F is entirely defined and continuous. The map F acts as an affine map on each maximal subset. In §26.4 we explain how we verify the following experimentally discovered facts.

- $\mathcal{X}[1/4,1/2]$ is partitioned into 19 maximal domains, each of which is a convex rational polytope. The vertices in the partition of $\mathcal{X}$ are of the form $(a/q,b/q,p/q)$ where $a,b,p,q\in\boldsymbol{Z}$ and

$$(p,q) \in \{(1,4),(2,7),(3,10),(1,3),(1,2)\}.$$

These polyhedra are permuted by the map $\iota_1(x,y,s) = (-x,-y,s)$. Note that p/q never lies in $(1/3,1/2)$.

- $\mathcal{X}[1/2,1]$ is partitioned into 13 maximal domains, each of which is a convex rational polytope. The vertices in the partition of $\mathcal{X}[1/2,1]$ are of the form $(a/q, b/q, p/q)$ where $a,b,p,q \in \boldsymbol{Z}$ and
$$(p,q) \in \{(1,2),(2,3),(3,4),(1,1)\}.$$
These polyhedra are permuted by ι_1.
- The 19 polyhedra in the partition of $\mathcal{X}[1,2]$ are all images of the polyhedra in the partition of $\mathcal{X}[1/4,1/2]$ under the map $\iota_2 \circ F$, where
$$\iota_2(x,y,s) \to ((x+y)/2s, (x-y)/2s, 1/2s).$$
These polyhedra are permuted by ι_1. This is a consequence of the Inversion Lemma from §7.5.

In §28.7 we will explain the action of the map F on each polyhedron. We scale all the polyhedra by a factor of $420 = 10 \times 7 \times 6$ so that we can make all our calculations using integer arithmetic, One can think of this rescaling as a way of clearing all denominators in advance of the calculations.

We will list the 420-scaled polyhedra in §28.6. For now, call them
$$\alpha_0, ..., \alpha_{18}, \beta_0, ..., \beta_{12}, \gamma_0, ..., \gamma_{18}.$$
We have labeled the list so that
$$\alpha_i \subset \mathcal{X}[105,210], \quad \beta_i \subset \mathcal{X}[210,420], \quad \gamma_i \subset \mathcal{X}[420,840]. \tag{26.4}$$
$\mathcal{X}[105,210]$ is our name for the 420-scaled version of $\mathcal{X}[1/4,1/2]$, etc. When we want to discuss these polyhedra all at once, we call them $P_0, ..., P_{50}$.

26.2. Avoiding Computational Error

The only operations we perform on vectors are vector addition and subtraction, the dot and cross product, scaling a vector by an integer, and dividing a vector by $d \in \boldsymbol{Z}$ provided d divides all the coordinates. These operations in turn only use plus, minus, and times, and integer division.

64-bit integer arithmetic: For Calculations 1-8, we we represent integers as *longs*, a 64 bit integer data type. One can represent any integer strictly between -2^{-63} and 2^{63}. (The extra bit gives the sign of the number.) There are two possible sources of error: overflow error and division errors.

Overflow: We subject all our calculations to an overflow checker, to make sure that the computer never attempts a basic operation (plus, minus, times) in with either the inputs or the output is out of range. To give an example, if we want to take the cross product $V_1 \times V_2$, we first check that all entries in V_1 and V_2 are less than 2^{30} in size. This guarantees that all intermediate answers, as well as the final answer, will be in the legal range for longs. The code is written so that the overflow checker halts the program (by throwing an exception) if some long is too large. This never happens when we actually run the calculations.

Division: We also check our division operations. Before we compute n/d, we make sure that $n \equiv 0 \bmod d$. The java operation $n\%d$ does this. Once we know that $n\%d = 0$, we know that the computer correctly computes the integer n/d.

BigIntegers: As we will discuss below, one part of Calculation 12 requires integers that overflow 64-bit arithmetic. For these calculations, we use the BigInteger class in Java. With the BigInteger class, one can perform addition, subtraction on integers of arbitrary length. (We never need to perform division in these calculations.) Here "arbitrary" means "subject to the limitations of the computer", but in practice one can use integers that have thousands of digits. Our calculations come nowhere near these limits.

26.3. Dealing with Polyhedra

Here we explain the methods we use when dealing with polyhedra.

No Collinearities: Given a polyhedron P, let $P_1, ..., P_n$ denote the vertices. We first check that

$$(P_k - P_i) \times (P_j - P_i) \neq 0, \qquad \forall\, i < j < k \in \{1, ..., n\}. \tag{26.5}$$

This guarantees that no three points in our vertex list of P are collinear. We found the polyhedra of interest to us in an experimental way, and initially they had many such collinearities. We detected collinearities by the failure of Equation 26.5, and then removed all the redundant points.

Face Lists: For each of our polyhedra P, we find and then store the list of faces of the polyhedron. To do this, we consider each subset $S = \{S_1, ..., S_m\}$ having at least 3 members. We check for three things.

(1) S lies in a single plane. We compute a normal $N = (S_2 - S_1) \times (S_3 - S_1)$ and then check that $N \cdot S_i$ is independent of i. Assuming this holds, let $D = N \cdot S_i$.
(2) S lies on ∂P. To check this, we compute the normal N as above, and then check that either $\max N \cdot P_i \leq D$ or $\min N \cdot P_i \geq D$.
(3) We check that S is maximal with respect to sets satisfying the first two properties.

Improved Normals: We noticed computationally that all of the normals to all of the polyhedron faces can be scaled so that they have the following form: At least two of the three coordinates lie in $\{-1, 0, 1\}$ and the third coordinate lies in $\{-8, ..., 8\}$. When we use the normals in practice, we make this scaling. This is one more safeguard against overflow error.

No Face Redundancies: Once we have the face list, we check that each vertex of each polyhedron lies in exactly 3 faces. In particular, all the vertices of our polyhedra are genuine vertices.

Containment Algorithm: Suppose we want to check if a vector V lies in a polyhedron P. For each face S of P, we let N be the (scaled) normal to S. we set $D = N \cdot S_0$, and then we verify the following.

- If $\max N \cdot P_i \leq D$ then $V \cdot N \leq D$.
- If $\min N \cdot P_i \geq D$ then $V \cdot N \geq D$.

If this always holds then V lies on the same side of S as does P, for all faces S. In this case, we know that $V \in P$. If we want to check that $V \in \text{interior}(P)$, we make the same tests, except that we require strict inequalities.

Disjointness Algorithm: Let $\boldsymbol{Z}_{10} = \{-10, ..., 10\}$. To prove that two polyhedra P and Q have disjoint interiors, we produce (after doing a search) an integer vector $W \in \boldsymbol{Z}_{10}^3$ such that

$$\max W \cdot P_i \leq \min W \cdot Q_j. \tag{26.6}$$

Volumes: For many of the polytopes P we consider, we compute 6 volume$(P) \in \boldsymbol{Z}$. To compute this volume, we decompose P into prisms by choosing a vertex of $v \in P$ and then computing

$$\sum_{f \in P} 6\text{volume}([v, f]). \tag{26.7}$$

The sum is taken over all faces f of P and $[v, f]$ denotes the prism obtained by taking the convex hull of $v \cup f$.

We arrange things so that $[v, f]$ is either is a tetrahedron or a pyramid with quadrilateral base. In the tetrahedron case,

$$6 \text{ vol}([v, f] = \det(M(v, f)), \tag{26.8}$$

where m is the matrix of vectors pointing from v to vertices of f. In the pyramid case,

$$6 \text{ vol}([v, f]) = \frac{1}{2} \sum_{i=1}^{4} \det(M(v, f_i) \tag{26.9}$$

where f_i is obtained from f by omitting the ith vertex.

Remark: In Calculation 12, we will compute volumes with BigIntegers. We will compute 12 vol$([v, f])$ instead, because we don't want to divide by 1/2.

26.4. Verifying the Partition

Now we explain how we verify that the polyhedra we work with are really correct. Let $\mathcal{RX}$ denote the polytope obtained from $\mathcal{X}$ by rotating 90 degrees. We have $F = (F')^2$, where F' is the original PET which swaps $\mathcal{X}$ and $\mathcal{RX}$.

Pairwise Disjointness: Using the Disjointness Algorithm, we check that P_i and P_j have disjoint interiors for all $i \neq j \in \{0, ..., 50\}$. We check the same thing for $F(P_i)$ and $F(P_j)$.

Containment: Using the Containment Algorithm, we check that

- $P_i \subset \mathcal{X}$ for $i = 0, ..., 50$.
- $F(P_i) \subset \mathcal{X}$ for $i = 0, ..., 50$.
- $F'(P_i) \subset \mathcal{R}X$ for $i = 0, ..., 50$.

We also see, by inspection, that F has a different action on P_i and P_j whenever P_i and P_j share a (non-horizontal) face. These checks show that each P_i is a maximal domain for the action of F.

Filling: It remains to check we check that $\mathcal{X}$ is partitioned into $P_0, ..., P_{50}$. We check that

$$\sum_{i=0}^{50} \text{volume}(P_i) = \text{volume}(\mathcal{X}). \tag{26.10}$$

The same equation shows that $\mathcal{X}$ is also partitioned into $F(P_0), ..., F(P_{50})$.

26.5. Verifying Outer Billiards Orbits

So far, we have been discussing the methods for Calculations 1-8. We use similar ideas for Calculations 9-12. Here we describe the main techniques.

For Calculations 9-12, the parameter interval is $[1/2, 1]$. Define

$$\mathcal{Y} = \boldsymbol{R}^2 \times [1/2, 1], \qquad \mathcal{Y}_y = \boldsymbol{R}^2 \times \{s\} \tag{26.11}$$

Let $P \subset \mathcal{Y}$ be a closed convex polyhedron. Let P_s denote the interior of $P \cap \mathcal{Y}_s$. We call P an *N-good polyhedron* if the first N iterates of the outer billiards map relative to any $s \in (1/2, 1)$ is defined and continuous on each slice P_s. In this section we explain how to verify that a polyhedron is N-good.

Figure 26.1 shows the partition for the outer billiards map ψ'. Let $v_1, ..., v_8$ be the vertices of O_s. We label the unbounded sectors $S_1, ..., S_8$ in such a way that v_j is the apex of S_j.

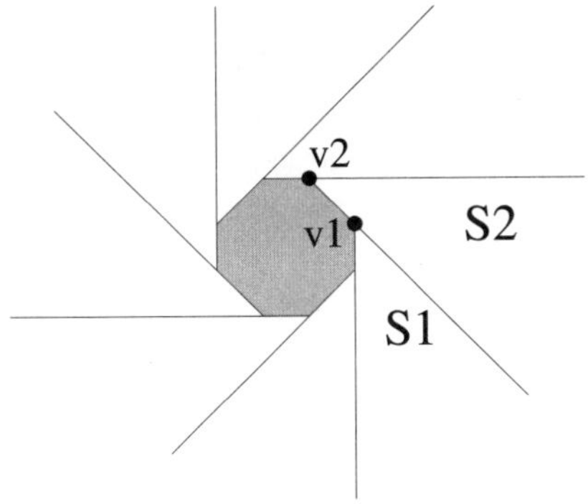

Figure 26.1: The partition for outer billiards.

Let $\mathcal{S}_j \subset \mathcal{Y}$ denote the union of points of the form (p, s) where $p \in S_j$ at the parameter s. We again call $\mathcal{S}_j$ a sector.

LEMMA 26.2. *The sectors are all closed convex subsets.*

Proof: Consider S_1. At the parameter s, the sector S_1 lies to to the right of the line $x = s$ and below the line $x + y = 1$. As s varies, these lines sweep out strips – i.e. subsets of planes – in $\mathcal{Y}$. The same goes for the other sectors. $\square$

The outer billiards map acts fiberwise on $\mathcal{Y}$. When restricted to the interior of each sector $\mathcal{S}_j$, the map is an affine involution which fixes the line segment swept out by the vertex v_j. Let Θ denote the outer billiards map as it acts on $\mathcal{Y}$. To say that a polyhedron P is N-good is to say that the first N iterates of Θ are defined on all points in the interior of P.

LEMMA 26.3. *Suppose that P is a closed convex polyhedron in $\mathcal{Y}$ and the vertices of P lie in the closed sector $\mathcal{S}_j$. Suppose also that P is not contained in a single plane. Then Θ is defined on all points in the interior of P.*

Proof: All points in the interior of P lie in the interior of $\mathcal{S}_j$, and Θ is defined on the interior of $\mathcal{S}_j$. □

The restriction of Θ to the interior of each sector $\mathcal{S}_j$ extends to the boundary, because the map in question is just the restriction of an affine map. We denote this extension by Θ_j. Note that Θ is not defined on points on the boundary of $\mathcal{S}_j$ because such points also belong to the boundaries of other sectors.

We say that the sequence $\{j\}$ is a *feasible sequence* for a polyhedron P_0 if all the vertices of P_0 lie in $\mathcal{S}_j$. In this case, we may define $P_1 = \Theta_j(P_0)$. Inductively, we say that the length $n+1$ sequence $\{j_0, ..., j_n\}$ is a *feasible sequence* for P_0 if $\{j_1, ..., j_n\}$ is a length n feasubility sequence for the image $P_1 = \Theta_{j_0}(P_0)$.

LEMMA 26.4. *Suppose that $P \subset \mathcal{Y}$ is a closed convex polyhedron,not contained in a single plane. If P has a length N feasible sequence then P is N-good. Moreover, the feasible sequence describes the successive sectors visited by the orbit in the interior of any point of P.*

Proof: This is an immediate consequence of Lemma 26.3 and induction. □

Remark: For the purposes of rigorous verification, it doesn't matter how we find the feasible sequences for the polyhedra we consider. In practice, however, we simply look at the orbit of a randomly chosen point in the interior. Once we have the candidate sequence, we verify that it works.

26.6. A Planar Approach

It might appear, from the form of Lemma 26.4. that somehow we will be doing 3-dimensional calculations. This is not true. We really just need to make a planar calculation for each vertex of the polyhedron.

Suppose that s is some fixed parameter and $p \in \boldsymbol{R}^2 - O_s$ is some point. We define feasible sequences for p just as we did in 3-dimensions. To say that the polyhedron P has a feasible sequence of length N is to say that there is one sequence σ of length N such that, for each vertex $p = (x, y, s)$ of P, the sequence σ is a feasible sequence for (x, y) at the parameter s.

Now we are going to explain how to do all the calculations using integer arithmetic. In a word, we rescale. We define a map T_m which simultaneously acts on $\boldsymbol{R}^2$ and on $\boldsymbol{R}^3$. We define T_m by the action

$$T_m(x, y) = (mx, my), \qquad T_m(x, y, s) = (mx, my, s). \tag{26.12}$$

When $s \in [1/2, 1)$, the vertices of O_s are not integers, and also the vertices of the polyhedra we test might not be integers. Since we want to make integer calculations, we will replace P and O_s by $T_m(P)$ and $T_m(O_s)$, where m is chosen so that the first two coordinates of all the vertices of $T_m(P)$ are integers and the coordinates of the vertices of $T_m(O_s)$ are integers. Once we have scaled this way, the check that some sequence is a feasible sequence for some point is an integer calculation.

It would nice if we could, as above, take $m = 420$ for all our calculations, but unfortunately we need to take much larger scale factors sometimes. The reason is that Calculations 11 and 12 involve the square of the octagonal PET, and so we need to work with the partition $\mathcal{A}_2$ rather than the simpler partition $\mathcal{A}_1$. Indeed, for Calculation 12, we will need to consider a few pieces from the partition $\mathcal{A}_4$. The vertices of the polyhedra in these partitions have rational coordinates with large denominators, and we need to rescale by fairly large integers to kill those denominators.

For the most part, these large integer scale factors are not a problem for us. However, there is one place in Calculation 12 where we need to compute the volumes of some of the scaled copies of our polyhedra coming from $\mathcal{A}_4$. For these volumes, the integers get slightly too large for te 64-bit arithmetic offered by the *longs*. So, for all the volume calculations associated to Calculations 9-12, whether they need it or not, we use the BigInteger class in java, which allows for integer calculations involving thousands of digits.

There is one more remark about volume we need to make. It turns out that our method of computing 6 times the volume, described above, involves dividing some even integers by 2. We never get division errors when we do such calculations with *longs* but with the Big Integers we would prefer to stay away from division altogether. So, we compute 12 times the volume instead.

26.7. Generating the Partitions

As we mentioned in the last section, Calculations 11 and 12 involve some polyhedra in the partitions $\mathcal{A}_2$ and $\mathcal{A}_4$. Once again, from the point of view of rigorous calculation, it doesn't matter how we produce these partitions. However, it seems worthwhile to say what we do.

For the parameter interval $[1/2, 1]$, there are 13 pieces in the partition $\mathcal{A}_1$, namely (unscaled copies of) $\beta_0, ..., \beta_{12}$ listed above. For each pair of indices (i, j) we compute the intersection

$$T_m(f(\beta_i)) \cap T_m(\beta_j). \tag{26.13}$$

where m is chosen so large that all the intersection points are integers.

To intersect two polyhedra P and Q, we consider the following 4 collections of vertices:

- All the vertices of P which are contained in Q.
- All the vertices of Q which are contained in P.
- All intersections of the form $A_1 \cap A_2 \cap B$, where A_1 and A_2 are faces of P and B is a face of Q.
- All intersections of the form $B_1 \cap B_2 \cap A$, where B_1 and B_2 are faces of P and A is a face of Q.

We then weed out redundant vertices – i.e. those which are not extreme points of the convex hull. All these operations are done using integer linear algebra.

We do something similar to get the few pieces of $\mathcal{A}_4$ that we need in Calculation 12.

We list all the vertices of all the polyhedra we use in the files for our program BonePet. We also have written the code which generates these polyhedra, and the reader can regenerate the data files directly from BonePet.

CHAPTER 27

The Calculations

27.1. Calculation 1

Recall that F is the octagonal PET map acting on $\mathcal{X}[1/4,1]$. Let $P_0,...,P_{31}$ be the polyhedra in the partition for F. Let $H = F^{-1}$. Then the polyhedra $F(P_0),...,F(P_{31})$ give a partition $\mathcal{H}$ for the map H. We rename the members of $\mathcal{H}$ as $H_0,...,H_{31}$.

Recall that we have the partition

$$X^0 = A \cup B, \tag{27.1}$$

where A and B are the symmetric pieces. Just for this calculation, we find it convenient to introduce the new symmetric piece

$$A^* = A \cup \text{central tiles} \cup \iota(A). \tag{27.2}$$

Here A^* is a hexagon which, like A, is symmetric under the reflection in the x-axis. We have a partition of X into 3 symmetric pieces

$$X = A^* \cup B \cup C, \qquad C = \iota(B). \tag{27.3}$$

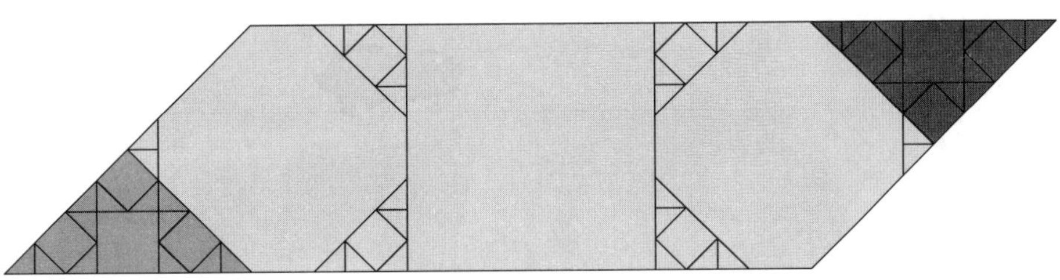

Figure 27.1: A^* (white) and B (light) and C (dark) for $s = 4/13$.

Let $\mathcal{A}$ be the polyhedron in $calX[1/4,1]$ which intersects the s-fiber in the set A^*_s. Likewise define $\mathcal{B}$ and $\mathcal{C}$.

We have the partition

$$\mathcal{X}[1/4,1] = \mathcal{A}[1/4,1] \cup \mathcal{B}[1/4,1] \cup \mathcal{C}[1/4,1] \tag{27.4}$$

Here $\mathcal{A}$ is such that the fiber of $\mathcal{A}$ over s is the hexagon A_s. The polyhedron $\mathcal{B}$ has the same definition relative to the triangle β_s. The polyhedron $\mathcal{C}$ is obtained from $\mathcal{B}$ by reflecting in the line $x = y = 0$.

The map μ from Calculation 1 acts on each of $\mathcal{A}$ and $\mathcal{B}$ and $\mathcal{C}$ as a reflection. We verify that each polyhedron α_i and β_i is a subset of one of these 3 big pieces.

Thus, μ acts on each polyhedron as a reflection. The new partition

$$\mathcal{X}[1/4,1] = \bigcup_{i=0}^{18} \mu(\alpha_i) \cup \bigcup_{i=0}^{12} \mu(\beta_i) \tag{27.5}$$

is the partition for the map

$$G = \mu \circ f \circ \mu^{-1}. \tag{27.6}$$

We call this partition $\mathcal{G}$, and we rename its members $G_0, ..., G_{31}$.

So, in summary, $\mathcal{G}$ is the partition for G and $\mathcal{H}$ is the partition for H. Next, we find a list of 48 pairs i, j so that

$$\text{interior}(G_i) \cap \text{interior}(H_j) \neq \emptyset$$

only if (i,j) lies on our list. More precisely, we use the Separation Algorithm to show that all other pairs have disjoint interiors.

Finally, we consider the grid

$$\Gamma = \{(20i, 20k, 105 + 10k) \;\; i = -42, ..., 42, \;\; j = -21, ..., 21, \;\; k = 0, ..., 31\}. \tag{27.7}$$

We check the identity $G = H$ on each point of Γ and we also check that at least one point of Γ is contained in each intersection $G_i \cap H_j$ for each of our 48 pairs. This suffices to establish the identity on all of $\mathcal{X}[1/4,1]$.

27.2. Calculation 2

Calculation 2 follows the same scheme as Calculation 1. Here we just explain the differences in the calculation.

- We set $H = F$ and $G = \nu F^{-1} \nu$.
- $\mathcal{H}$ is the partition consisting of $\alpha_0, ..., \alpha_{18}$.
- $\mathcal{X}[1/4, 1/2]$ is partitioned into 5 smaller polyhedra, coming from P_s, Q_s, the central tile, $\iota(P_s)$ and $\iota(Q_s)$. the map ν acts as a reflection on each piece. For $i = 0, ..., 18$, the polyhedron $F(\alpha_i)$ is contained in one of the 5 pieces, so that ν acts isometrically on $F(\alpha_i)$.
- $\mathcal{G}$ be the partition of $\mathcal{X}[1/4, 1/2]$ by the polyhedra

$$\nu \circ F(\alpha_0), ..., \nu \circ F(\alpha_{18}).$$

- We find a list of 27 pairs (i,j) such that G_i and H_j do not have disjoint interiors.

The rest of the calculation is the same.

27.3. Calculation 3

Let $s \in [5/4, 2]$ and let $t = s - 1 \in [1/4, 1]$. We want to show that ϕ_s conjugates $f_t|Y_t$ to $f_s|Z_s$ and that every orbit of f_s intersects Z_s, except the following orbits.

- Those in the trivial tile $(\alpha_0 \cup \beta_0)_s$ of Δ_s.
- Those in the set

$$\tau_s = \phi_s\Big((\alpha_0 \cup \beta_0)_t\Big).$$

Once we are done, we will know that τ_s is in fact a tile of Δ_s, and that τ_s has period 2.

For this section we set $\mathcal{X} = \mathcal{X}[5/4, 2]$. Let $\mathcal{Y}$ denote the subset of $\mathcal{X}$ whose fiber over s is the set Y_s. Define $\mathcal{Z}$ in a similar way. The maps $\phi_s : Y_t \to Z_s$ piece together to give an isometry $\phi : \mathcal{Y} \to \mathcal{Z}$. The map is given by

$$\phi(x, y, z) = (x \pm 1, y \pm 1, z - 1) \tag{27.8}$$

Whether we add or subtract 1 to the first two coordinates depends on whether the point (x, y, z) lies in the left half of $\mathcal{Y}$ or in the right half.

For what we describe next, we always refer to open polyhedra, and our equalities are meant to hold up to sets of codimension 1, namely the boundaries of our polyhedra.

We have

$$\mathcal{Y} = \alpha_1 \cup ... \cup \alpha_{18} \cup \beta_1 \cup ... \cup \beta_{12}. \tag{27.9}$$

For each $i = 1, ..., 18$ we check computationally that there is some $k = k_i$ with the following three properties.

(1) The first $k_i + 1$ iterates of F^{-1} are defined on $\phi \circ F(\alpha_i)$. This amounts to checking that

$$P_{ij} = F^{-i} \circ \phi \circ F(\alpha_i) \tag{27.10}$$

is contained in some α_a or β_b for suitable indices a and b, and for all $i = 0, ..., k_i$.

(2)

$$P_{ij} \cap \mathcal{Z} = \emptyset, \qquad j = 1, ..., k_i. \tag{27.11}$$

Equation 27.11 shows that

$$F|P_{i0} = F^{k_i}$$

That is, on P_{i0}, the map F returns to $\mathcal{Z}$ as F^{k_i}. To establish Equation 27.11, we use the Separation Algorithm so show that

$$P_{ij} \cap \phi(\alpha_a) = \emptyset, \qquad P_{ij} \cap \phi(\beta_b) = \emptyset$$

for all $a = 1, ..., 18$ and $b = 1, ..., 12$, and all relevant indices i and j. This suffices because $\mathcal{Z}$ is partitioned into the polyhedra

$$\mathcal{Z} = \phi(\alpha_1) \cup ... \cup \phi(\alpha_{18}) \cup \phi(\beta_1) \cup ... \cup \phi(\beta_{12}).$$

(3) $P_{i,k_i+1} = \phi(\alpha_i)$.

We make all the same calculations for $\beta_1, ..., \beta_{12}$, finding an integer ℓ_i which works for β_i. We define Q_{ij} with respect to β_i just as we defined P_{ij} with respect to α_i.

Our calculations above show that ϕ conjugates $F|\mathcal{Y}$ to $F|\mathcal{Z}$. Also, by construction, the boundary of $\mathcal{Z}$ is contained in the union of the boundaries of the polyhedra $\phi(\alpha_i) \cup \phi(\beta_j)$. Hence, Z_s is a clean set for all $s \in [5/4, 2]$.

We still want to see that all orbits except those of period 1 and 2 actually intersect $\mathcal{Z}$. We check the following:

(1) F is entirely defined on $\phi(\alpha_0)$ and has order 2.
(2) Both $\phi(\alpha_0)$ and $F \circ \phi(\alpha_0)$ are disjoint from $\mathcal{Z}$. We use the same trick with the Separation Algorithm to do this.

We claim that the open polyhedra in the following union are pairwise disjoint.

$$\bigcup_{i=1}^{18}\bigcup_{j=0}^{k_i} P_{ij} \cup \bigcup_{i=1}^{12}\bigcup_{i=0}^{\ell_i} Q_{ij} \cup \bigcup_{j=0}^{1} F^j \circ \phi(\alpha_0) \cup \bigcup_{j=0}^{1} F^j \circ \phi(\beta_0) \cup \alpha_0 \cup \beta_0. \tag{27.12}$$

Suppose, for instance, that P_{ab} and P_{cd} were not disjoint. Then $P_{a,b+e}$ and $P_{c,d+f}$ would not be disjoint for $e > 0$ and $f > 0$ such that $b+e = k_a+1$ and $c+f = k_b+1$. But we know that these last polyhedra are disjoint because they respectively equal the disjoint polyhedra $\phi(\alpha_a)$ and $\phi(\alpha_c)$. Similar arguments work for the other cases.

Similar to Equation 26.10, we compute the sum of the volumes of the polyhedra in Equation 27.12 and see that it coincides with the volume of $\mathcal{X}$. Thus, $\mathcal{X}$ is partitioned into the polyhedra in Equation 27.12. This fact implies the all orbits except those of period 1 and 2 actually intersect $\mathcal{Z}$.

Finally, we see by process of elimination that τ_s really is a tile of Δ_s. All other points not in the interior of τ_s either have undefined orbits, or lie in the trivial tile, or have orbits which intersect $\mathcal{Z}$. Thus f_s cannot be defined on any point of the boundary of τ_s. Since f_s is defined, and has period 2, on the interior of τ_s, we see that τ_s is a tile of Δ_s having period 2.

27.4. Calculation 4

Calculation 4 follows the same scheme as Calculation 3. Here we will describe the differences between the two calculations.

- We consider the behavior of polyhedra on the interval $s \in [1/2, 3/4]$ rather than on $[5/4, 1]$. Here $t = 1 - s \in [1/4, 1/2]$.
- The map ϕ is not an isometry here, but rather a volume preserving affine map. The formula is

$$\phi(x, y, z) = (x \pm (1 - 2z), y \pm (1 - 2z), 1 - z). \tag{27.13}$$

  The choice of plus or minus again depends on whether (x, y, z) lies in the left of the right half of $\mathcal{Y}$.
- $\mathcal{Y}$ is partitioned into the tiles $\alpha_1, ..., \alpha_{18}$. The B-tiles are not needed here.
- The tiles τ and $\iota(\tau)$ already belong to $\mathcal{Z}$. The work in Calculation 3 shows that τ_s and ι_s are indeed period 2 tiles of Δ_s. This time, τ and $\iota(\tau)$ are amongst the images of $\alpha_1, ..., \alpha_{18}$ under ϕ.
- Using the notation from the previous section, the partition in Equation 27.12 becomes

$$\bigcup_{i=1}^{18}\bigcup_{j=0}^{k_i} P_{ij} \cup \alpha_0 \tag{27.14}$$

The rest of the calculation is the same.

27.5. Calculation 5

Calculation 5 follows the same scheme as Calculation 3, except that we don't need to keep track of the volumes. Let T and ω be as in Calculation 5. Let $s \in (1, 4/3]$ and let $t = T(s) \in (1, 2]$.

We define $\mathcal{W}$ and $\mathcal{Y}$ as the global versions of W_s and Y_u, as in Calculation 3. We are interested in $\mathcal{Y}[1, 2]$ and $\mathcal{W}[1, 4/3]$. Similar to Calculation 3, we have a global map $\omega : \mathcal{Y} \to \mathcal{W}$. We have the formula

$$\omega(x,y,z) = \big(\omega_s(x,y), s\big), \qquad s = T^{-1}(z). \tag{27.15}$$

We want to see that ω conjugates $F|\mathcal{Y}$ to $F|\mathcal{W}$.

We have

$$\mathcal{Y} = \gamma_1 \cup \ldots \cup \gamma_{18}. \tag{27.16}$$

By the same methods used in Calculation 3, we check, for each $i = 1, \ldots, 18$, that there is some $k = k_i$ with the following three properties.

(1) The first $k_i + 1$ iterates of F^{-1} are defined on $\omega \circ F(\gamma_i)$. Define

$$P_{ij} = F^{-i} \circ \omega \circ F(\gamma_i) \tag{27.17}$$

(2)

$$P_{ij} \cap \mathcal{W} = \emptyset, \qquad j = 1, \ldots, k_i. \tag{27.18}$$

(3) $P_{i,k_i+1} = \omega(\alpha_i)$.

These facts imply that ω conjugates $F|\mathcal{Y}$ to $F|\mathcal{W}$.

Finally, the set Z_s is clean for each s for the following reasons.

- The top edge of Z_s^0 and the bottom edge of $\iota(Z_s^0)$ are contained in the union of slices of the sets $\omega \circ F(\partial\gamma_i)$.
- The vertical edges of Z_s are contained in the set $\partial\tau_s \cup \iota(\partial\tau_s)$.
- The remaining edges of Z_s lie in the ∂X_s.

27.6. Calculation 6

As we mentioned in §5, Calculation 6 practically amounts to inspecting the partition. For Statement 1, we let τ be the polyhedron which restricts to τ_s for $s \in [1, 3/2]$. We list this polyhedron in §28.2. We check that F is entirely defined on (the interior of) τ and that $F^2(\tau) = \tau$.

For each polyhedron P, let P_s denote the intersection of P with the horizontal plane of height s.

For Statement 2, let $\mathcal{Z}$ be the polyhedron which restricts to Z_s^0 for $s \in [1, 5/4]$. We compute that

$$\mathcal{Z} \subset F(\gamma_{13}). \tag{27.19}$$

We also try a single point $(x, y, s) \in F(\gamma_{13})$ and check that $f_s^{-1}(p) = p + \delta_s$. Since $F(\gamma_{13})$ is a domain of continuity for F^{-1}, the same result holds for all points in $F(\gamma_{13})$, including all the points in $\mathcal{Z}$. This proves Statement 2.

For Statement 3, let $\mathcal{K}$ be the union of two polyhedra which intersect the fiber X_s in $X_s - Z_s - W_s$, for $s = (1, 5/4]$. We see by inspection that

$$\mathcal{K} = F(\gamma_2) \cup F(\gamma_8) \cup F(\gamma_{11}) \cup F(\gamma_{17}); \qquad \gamma_j \subset \mathcal{Z}, \qquad j = 2, 8, 11, 17. \tag{27.20}$$

This proves Statement 3.

Remark: We could have made an explicit computation to establish Equation 27.20, but this is something that is obvious from a glance at just 2 planar pictures. We just have to check Equation 27.20 at the parameters $s = 1$ and $s = 3/2$ because every polyhedron P in sight, when restricted to the fibers above $[1, 5/4]$, is the convex hull of $P_1 \cup P_{5/4}$.

27.7. Calculation 7

Calculation 7 follows the same scheme as Calculation 5. Here are the differences.

- Here we are interested in $\mathcal{Y}[1/2,1]$ and $\mathcal{W}[3/4,1]$.
- Here we use the formula from 10.7 to define tha map ω in Equation 27.15.
- Here we have

$$\mathcal{Y} = \beta_1 \cup ... \cup \beta_{12}. \tag{27.21}$$

The rest of the calculation is the same.

27.8. Calculation 8

Calculation 8 works essentially the same was as Calculation 6. For Statement 1, we let τ be the polyhedron which restricts to τ_s for $s \in [3/4,1]$. We list this polyhedron in §28.2. We check that F is entirely defined on (the interior of) τ and that $F^2(\tau) = \tau$.

For Statement 2, we let $\mathcal{Z}^*$ be the polyhedron which intersects X_s in $(Z_s^0)^*$ for $s \in [3/4,1)$. We compute that

$$\mathcal{Z}^* \subset \beta_7 \tag{27.22}$$

and we finish the proof of Statement 2 just as in Calcultion 6.

For Statement 3, we define $\mathcal{K}$ as in Calculation 6 and we see by inspection that

$$\mathcal{K} = \beta_2 \cup \beta_6 \cup \beta_8 \cup \beta_{12}, \qquad F(\beta_j) \subset \mathcal{Z}, \qquad j = 2,6,8,12. \tag{27.23}$$

This proves Statement 3.

27.9. Calculation 9

Calculation 10 contains Calculation 9, so we will only describe Calculation 10.

27.10. Calculation 10

Consider the polyhedron O having vertices

$$(\pm 1/2, \pm 1/2, 1/2), \qquad (\pm 1, 0, 1), \qquad (0, \pm 1, 1). \tag{27.24}$$

O has the property that $O \cap \mathcal{Y}_s = O_s$, the central octagon.

We want to verify that the translate $O_s + (n,0)$ is a periodic tile, at least in case $n = 2,4,6,8,10$. We consider the polyhedra

$$O(n) = O + (n,0,0), \qquad n = 2,4,6,8,10. \tag{27.25}$$

We want to show that the octagon $O(n)$ has period $8n$. For each $O(n)$ we compute a feasibility sequence of length $8n$ and then verify that it is indeed a feasibility sequence. We then apply Lemma 26.4 in each of the 5 cases. Here are the 5 feasible sequences:

- 14725036
- 1504736251403726
- 151404737262515040373626
- 15150404737362625151404037372626
- 1515140404737372626251515040403737362626

Remark: Our calculation actually proves the weaker statement that the octagons of interest to us are contained in periodic tiles. However, the top of each octagon is contained in the boundary of one of the strips mentioned in §5.1. Hence, the top edge of each octagon is really a boundary of a periodic tile. The same goes for the bottom edge. Following the orbit around, we see that the same is true of all the edges. Hence, the octagons really are periodic tiles.

27.11. Calculation 11

For the purposes of this calculation, we think of the dogbone D as the domain for the octagonal PET map f. Let $\mathcal{A}_2$ be the partition for f^2, the square of the octagonal PET. So, $\mathcal{A}_2$ is a partition of D. As we discussed in the previous chapter, we rescale all these polyhedra so that the first two coordinates of every vertex of every polyhedron in the partition is an integer. The scale factor is 840 (just twice what we needed for Calculations 1-8.)

$\mathcal{A}_2$ has 40 polyhedra, $P_0, ..., P_{39}$. We list these in §28.8. For each such polyhedron P_j, we consider the translates

$$P_{jn} = P_j + (n, 0), \qquad n = 2, 4, 6, 8. \tag{27.26}$$

We first check that $\{P_{jn}\}$ for $j = 0, ..., 39$ forms a partition of the dogbone D_n^0. Really, we just have to check this for $n = 2$. The remaining cases follow from symmetry. Also, this check is somewhat redundant, on account of the way we created the partition $\mathcal{A}_2$. Nonetheless, we make the check by showing that the polyhedra are pairwise disjoint and that the sum of the volumes is as expected.

Next, we associate to P_{jn} a sequence σ_{jn} and also a number $\epsilon_{jn} \in \{0, 1\}$. The sequence is supposed to encode the orbit of P_{jn} until it returns to D_n as $\Psi(P_{jn})$. The number $\epsilon = \epsilon_{jn}$ is supposed to be such that

$$\Psi(P_{jn}) \subset D_n^{\epsilon}. \tag{27.27}$$

We verify that σ_{jn} is a feasible sequence for P_{jn} and then we compute the image P'_{jn} under the corresponding power of the outer billiards map Θ. We then verify that

- $f^2(v) - v = v'_n - v_n$ if $\epsilon_{jn} = 0$.
- $f^2(v) - v = \iota(v'_n) - v_n$ if $\epsilon_{jn} = 1$.

holds on all the vertices v of P_j. These symbols have the following meaning:

- $v_n = v + (n, 0)$ is the vertex on P_{jn} corresponding to v.
- v'_n is the vertex on P'_{jn} corresponding to v_n.
- $\iota : D_n \to D_n$ is the involution which swaps the dogbones D_n^0 and D_n^1.

Our calculations show that there is some exponent $\alpha = \alpha_{jn}$ so that Υ^α and f^2 agree on P_{jn}. In other words, we do not explicitly rule out the possibility that σ_{jn} describes some power of Ψ acting on P_{jn}. However, based on the length of σ_{jn} and the description of how the pieces circulate around the central octagon, we can see that σ_{jn} isn't long enough to describe the second return map, let along a higher power. One can also see this by directly inspecting the sequences. So Υ and f^2 agree on P_{jn} for each pair of indices (j, n).

27.12. Calculation 12

For this last calculation, we are trying to show that the map $p \to p + (2, 0)$ conjugates $\Psi|E_1$ to $\Psi|E_3$.

For starters, we note that we do not need to prove this result for all $s \in [1/2, 1]$. The two octahedra O_s and O_t are isometric when $s = 1/(2t)$. For this reason, it suffices to prove our result for $t \in [\sqrt{2}/2, 1]$. In fact, we use the parameter interval $[2/3, 1]$. We did not set out to make this restriction; we just found by trial and error that we could use simpler partitions if we restricted the parameter interval this way.

We produce partitions on E_1 and E_3 so that

- For polyhedron P_1 in the partition of E_1 there is a corresponding polyhedron P_3 in the partition of E_3 so that $P_3 = P_1 + (2, 0)$.
- The first return map to E_1 (of the square of the outer billiards map) is defined on each polyhedron in the partition of E_1.
- The first return map to E_3 (of the square of the outer billiards map) is defined on each polyhedron in the partition of E_3.
- Let P_1 and P_3 be corresponding polyhedra. Let P_1' and P_3' be the corresponding images under the first return map. We check that

$$v_1' - v_1 = v_3' - v_3 \tag{27.28}$$

  for all vertices v_1 of P_1. Here $v_3 = v_1 + (0, 2, 0)$. Also v_j is the vertex of P_j' corresponding to v_j.

Once we have the partitions, we verify all the statements in the same vertex-by-vertex way we do for Calculation 11.

We will describe the partition of E_1. The partition of E_3 is just obtained by translating each polyhedron. The partition is built in 3 steps.

Start with What We have: We start with the partition $\mathcal{P}_2$ consisting of polyhedra

$$P_j + (0, 2), \qquad j = 0, ..., 39. \tag{27.29}$$

These are the 40 polyhedra from $\mathcal{A}_2$ used in Calculation 11.

Extend by Symmetry: Next, there is a translation **up** (respectively **dn**) which maps the bottom (respectively top) half of D into the top (respectively bottom) quarter of E_1. See Figure 27.2 for the actions of these maps.

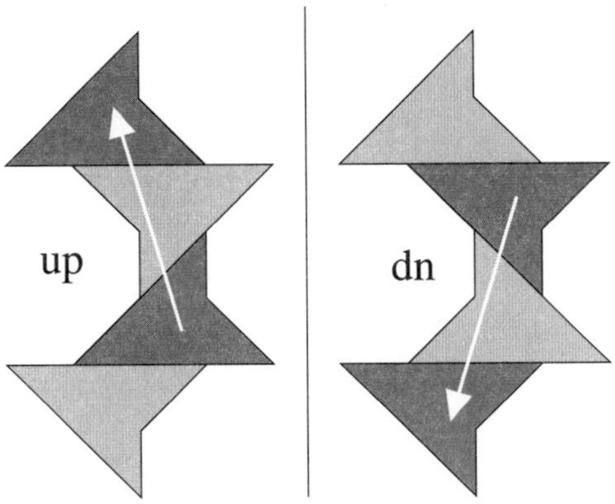

Figure 27.2: The action of j_1.

We pull back the relevant 20 pieces of $\mathcal{P}_2$ by j_1 and j_2 to get a partition $\mathcal{P}_3$ consisting of 80 pieces. This is a partition of E_1.

Subdivide When Necessary: It would be nice to simply say that we use the partition $\mathcal{P}_3$ for our calculation. However, it turns out (even restricting to the interval $[2/3, 1]$) that 4 out of 80 of the pieces fail the feasibility test. What is going is that the map $\Psi|E_1$ is not entirely defined on these pieces. In our listing the bad pieces are $P_j + (2, 0)$ for $j = 0, 28, 58, 67$.

We replace each bad piece $P_j + (2, 0)$ by the finite union of pieces

$$P_{jk} + (0, 2),$$

where the collection $\{P_{jk}\}$ is the portion of the partition $\mathcal{A}_4$ that is contained in P_j. It turns out that we are replacing 4 polyhedra with 25 polyhedra. So, the final partition $\mathcal{P}_4$ has 101 polyhedra in it. Our calculations work for this partition.

Remarks:
(i) A more robust method would be just to use translates of $\mathcal{A}_4$ in the first place, but this seemed to involve a massive calculation that would frequently overflow 64 bit arithmetic and require extensive use of the BigInteger class.
(ii) We might have tried to use polyhedra from the simpler partition $\mathcal{A}_3$ but we felt that probably it would not work. The even powers of the octagonal PET are well-related to the outer billiards systems we have been considering, but the odd ones do not seem well related.

CHAPTER 28

The Raw Data

28.1. A Guide to the Files

In this chapter, we will list enough data so that a very interested reader could reconstruct all our calculations from what is written here. However, our computer files also contain the same data, and more. All the computer files are contained in the two directories **OctaPET** and **BonePET**. We did not try to merge these two directories because the notation is not completely consistent. **OctaPET** involves the calculations (Calculations 1–8) related to properties of the octagonal PET, and **BonePET** involves the calculations (Calculations 9–12) involving outer billiards. Here is a guide to the files.

- The file **OctaPET/DataPartition.java** contains the listing of the polyhedra in our partition for the octagonal PET map.
- The file **OctaPET/DataPartitionExtra.java** contains the listing of the remaining polyhedra used in Calculations 1-8, such as those corresponding to the symmetric pieces A, B, P, Q.
- The file **BonePET/DataPartition1Raw.java** contains the listing of the polyhedra in our partition $\mathcal{A}_1$ for the octagonal PET defined on the dogbone D discussed in §5 and in §27.11. As we explain in §28.8, the 13 polyhedra listed here are just translates of 13 of the polyhedra listed in **OctaPET/DataPartition.java**.
- The file **BonePET/DataPartition2Raw.java** contains the listing of the polyhedra in the partitions $\mathcal{A}_2$ and $\mathcal{A}_3$ discussed in §27.12.
- The file **BonePET/DataPartition4Raw.java** contains the listing of the polyhedra in the partitions $\mathcal{A}_4$. §27.12.

Unless otherwise mentioned, all polyhedra are scaled by 420.

28.2. The Main Domain

Here is $\mathcal{X}[1/4, 2]$. This is the big polyhedron which contains all the others.

$$\begin{bmatrix}-525\\-105\\105\end{bmatrix}\begin{bmatrix}315\\-105\\105\end{bmatrix}\begin{bmatrix}-315\\105\\105\end{bmatrix}\begin{bmatrix}525\\105\\105\end{bmatrix}\begin{bmatrix}-1260\\-840\\840\end{bmatrix}\begin{bmatrix}-420\\-840\\840\end{bmatrix}\begin{bmatrix}420\\840\\840\end{bmatrix}\begin{bmatrix}1260\\840\\840\end{bmatrix}$$

28.3. The Symmetric Pieces

Here are the polyhedra corresponding to the symmetric pieces used in Calculations 1 and 2.

Here is $\mathcal{A}[1/4, 1]$. Note that this piece is based on A^* rather than A. See §27.1.

$$\begin{bmatrix}315\\105\\105\end{bmatrix}\begin{bmatrix}-315\\105\\105\end{bmatrix}\begin{bmatrix}-315\\-105\\105\end{bmatrix}\begin{bmatrix}315\\-105\\105\end{bmatrix}\begin{bmatrix}420\\0\\105\end{bmatrix}\begin{bmatrix}-420\\0\\105\end{bmatrix}\begin{bmatrix}0\\420\\420\end{bmatrix}\begin{bmatrix}420\\0\\420\end{bmatrix}\begin{bmatrix}-420\\0\\420\end{bmatrix}\begin{bmatrix}0\\-420\\420\end{bmatrix}$$

Here is $\mathcal{B}[1/4,1]$.

$$\begin{bmatrix}-315\\-105\\105\end{bmatrix}\begin{bmatrix}-420\\0\\105\end{bmatrix}\begin{bmatrix}-525\\-105\\105\end{bmatrix}\begin{bmatrix}-840\\-420\\420\end{bmatrix}\begin{bmatrix}-420\\0\\420\end{bmatrix}\begin{bmatrix}0\\-420\\420\end{bmatrix}$$

Here is $\mathcal{P}[1/4,1/2]$.

$$\begin{bmatrix}-315\\-105\\105\end{bmatrix}\begin{bmatrix}-105\\-105\\105\end{bmatrix}\begin{bmatrix}-105\\105\\105\end{bmatrix}\begin{bmatrix}-315\\105\\105\end{bmatrix}\begin{bmatrix}-630\\-210\\210\end{bmatrix}\begin{bmatrix}-210\\-210\\210\end{bmatrix}\begin{bmatrix}-210\\210\\210\end{bmatrix}$$

Here is $\mathcal{Q}[1/4,1/2]$.

$$\begin{bmatrix}-525\\-105\\105\end{bmatrix}\begin{bmatrix}-315\\-105\\105\end{bmatrix}\begin{bmatrix}-315\\105\\105\end{bmatrix}\begin{bmatrix}-630\\-210\\210\end{bmatrix}$$

28.4. Period Two Tiles

Here is the period 2 tile $\tau[1,5/4]$ from Calculation 6.

$$\begin{bmatrix}-525\\-315\\315\end{bmatrix}\begin{bmatrix}-315\\-315\\315\end{bmatrix}\begin{bmatrix}-315\\-105\\315\end{bmatrix}\begin{bmatrix}-525\\-105\\315\end{bmatrix}\begin{bmatrix}-420\\-420\\420\end{bmatrix}$$

Here is the period 2 tile $\tau[3/4,1]$ from Calculation 8.

$$\begin{bmatrix}-420\\-420\\420\end{bmatrix}\begin{bmatrix}-210\\-630\\630\end{bmatrix}\begin{bmatrix}-210\\-210\\630\end{bmatrix}\begin{bmatrix}-630\\-210\\630\end{bmatrix}\begin{bmatrix}-630\\-630\\630\end{bmatrix}$$

28.5. The Domains from the Main Theorem

Here is the domain $\mathcal{Z}[1,5/4]$ from Calculation 6.

$$\begin{bmatrix}-420\\-420\\420\end{bmatrix}\begin{bmatrix}-840\\-420\\420\end{bmatrix}\begin{bmatrix}-945\\-525\\525\end{bmatrix}\begin{bmatrix}-525\\-525\\525\end{bmatrix}\begin{bmatrix}-525\\-315\\525\end{bmatrix}\begin{bmatrix}-735\\-315\\525\end{bmatrix}$$

Here is the domain $\mathcal{Z}[3/4,1]$ from Calculation 8.

$$\begin{bmatrix}-420\\-420\\420\end{bmatrix}\begin{bmatrix}-840\\-420\\420\end{bmatrix}\begin{bmatrix}-945\\-525\\525\end{bmatrix}\begin{bmatrix}-525\\-525\\525\end{bmatrix}\begin{bmatrix}-525\\-315\\525\end{bmatrix}\begin{bmatrix}-735\\-315\\525\end{bmatrix}$$

28.6. The Polyhedra in the Partition

Define

$$\iota_1(x,y,s)=(-x,-y,s), \qquad \iota_2(x,y,s)=\left(\frac{x+y}{2s},\frac{x-y}{2s},\frac{1}{2s}\right). \tag{28.1}$$

The partition of $\mathcal{X}[1/4,1/2]$ consists of the 19 polyhedra.

$$\alpha_0,\alpha_1,...,\alpha_9,\iota_1(\alpha_1),...,\iota_1(\alpha_9). \tag{28.2}$$

The partition of $\mathcal{X}[1/2,1]$ consists of the 13 polyhedra

$$\beta_0,\beta_1,...,\beta_6,\iota_1(\beta_1),...,\iota_1(\beta_6). \tag{28.3}$$

The partition of $\mathcal{X}[1,2]$ consists of the 19 polyhedra

$$\iota_2 \circ F(\alpha_i), \qquad i = 0, ..., 18. \tag{28.4}$$

Here $\alpha_0, ..., \alpha_{18}$ are the polyhedra from Equation 28.2.

$$A0 = \begin{bmatrix}105\\105\\105\end{bmatrix} \begin{bmatrix}-105\\105\\105\end{bmatrix} \begin{bmatrix}-105\\-105\\105\end{bmatrix} \begin{bmatrix}105\\-105\\105\end{bmatrix} \begin{bmatrix}210\\210\\210\end{bmatrix} \begin{bmatrix}-210\\210\\210\end{bmatrix} \begin{bmatrix}-210\\-210\\210\end{bmatrix} \begin{bmatrix}210\\-210\\210\end{bmatrix}$$

$$A1 = \begin{bmatrix}420\\0\\105\end{bmatrix} \begin{bmatrix}525\\105\\105\end{bmatrix} \begin{bmatrix}315\\105\\105\end{bmatrix} \begin{bmatrix}280\\140\\140\end{bmatrix}$$

$$A2 = \begin{bmatrix}105\\105\\105\end{bmatrix} \begin{bmatrix}280\\140\\140\end{bmatrix} \begin{bmatrix}140\\140\\140\end{bmatrix} \begin{bmatrix}210\\-210\\210\end{bmatrix}$$

$$A3 = \begin{bmatrix}280\\140\\140\end{bmatrix} \begin{bmatrix}140\\140\\140\end{bmatrix} \begin{bmatrix}210\\210\\210\end{bmatrix} \begin{bmatrix}210\\-210\\210\end{bmatrix} \begin{bmatrix}420\\0\\210\end{bmatrix}$$

$$A4 = \begin{bmatrix}525\\105\\105\end{bmatrix} \begin{bmatrix}420\\0\\140\end{bmatrix} \begin{bmatrix}420\\0\\210\end{bmatrix} \begin{bmatrix}630\\210\\210\end{bmatrix} \begin{bmatrix}210\\210\\210\end{bmatrix}$$

$$A5 = \begin{bmatrix}315\\-105\\105\end{bmatrix} \begin{bmatrix}420\\0\\105\end{bmatrix} \begin{bmatrix}315\\105\\105\end{bmatrix} \begin{bmatrix}420\\0\\120\end{bmatrix} \begin{bmatrix}378\\-42\\126\end{bmatrix}$$

$$A6 = \begin{bmatrix}420\\0\\120\end{bmatrix} \begin{bmatrix}462\\42\\126\end{bmatrix} \begin{bmatrix}420\\0\\140\end{bmatrix} \begin{bmatrix}420\\140\\140\end{bmatrix} \begin{bmatrix}280\\140\\140\end{bmatrix} \begin{bmatrix}210\\210\\210\end{bmatrix}$$

$$A7 = \begin{bmatrix}315\\105\\105\end{bmatrix} \begin{bmatrix}420\\0\\120\end{bmatrix} \begin{bmatrix}378\\-42\\126\end{bmatrix} \begin{bmatrix}420\\0\\140\end{bmatrix}$$

$$A8 = \begin{bmatrix}315\\-105\\105\end{bmatrix} \begin{bmatrix}315\\105\\105\end{bmatrix} \begin{bmatrix}105\\105\\105\end{bmatrix} \begin{bmatrix}105\\-105\\105\end{bmatrix} \begin{bmatrix}420\\0\\140\end{bmatrix} \begin{bmatrix}280\\140\\140\end{bmatrix} \begin{bmatrix}210\\-210\\210\end{bmatrix} \begin{bmatrix}420\\0\\210\end{bmatrix}$$

$$A9 = \begin{bmatrix}420\\0\\105\end{bmatrix} \begin{bmatrix}525\\105\\105\end{bmatrix} \begin{bmatrix}420\\0\\120\end{bmatrix} \begin{bmatrix}462\\42\\126\end{bmatrix} \begin{bmatrix}420\\140\\140\end{bmatrix} \begin{bmatrix}280\\140\\140\end{bmatrix}$$

$$B0 = \begin{bmatrix}105\\105\\105\end{bmatrix} \begin{bmatrix}-105\\105\\105\end{bmatrix} \begin{bmatrix}-105\\-105\\105\end{bmatrix} \begin{bmatrix}105\\-105\\105\end{bmatrix} \begin{bmatrix}210\\210\\210\end{bmatrix} \begin{bmatrix}-210\\210\\210\end{bmatrix} \begin{bmatrix}-210\\-210\\210\end{bmatrix} \begin{bmatrix}210\\-210\\210\end{bmatrix}$$

$$B1 = \begin{bmatrix}420\\0\\105\end{bmatrix} \begin{bmatrix}525\\105\\105\end{bmatrix} \begin{bmatrix}315\\105\\105\end{bmatrix} \begin{bmatrix}280\\140\\140\end{bmatrix}$$

$$B2 = \begin{bmatrix}105\\105\\105\end{bmatrix} \begin{bmatrix}280\\140\\140\end{bmatrix} \begin{bmatrix}140\\140\\140\end{bmatrix} \begin{bmatrix}210\\-210\\210\end{bmatrix}$$

$$B3 = \begin{bmatrix}280\\140\\140\end{bmatrix} \begin{bmatrix}140\\140\\140\end{bmatrix} \begin{bmatrix}210\\210\\210\end{bmatrix} \begin{bmatrix}210\\-210\\210\end{bmatrix} \begin{bmatrix}420\\0\\210\end{bmatrix}$$

$$B4 = \begin{bmatrix}525\\105\\105\end{bmatrix} \begin{bmatrix}420\\0\\140\end{bmatrix} \begin{bmatrix}420\\0\\210\end{bmatrix} \begin{bmatrix}630\\210\\210\end{bmatrix} \begin{bmatrix}210\\210\\210\end{bmatrix}$$

$$B5 = \begin{bmatrix}315\\-105\\105\end{bmatrix} \begin{bmatrix}420\\0\\105\end{bmatrix} \begin{bmatrix}315\\105\\105\end{bmatrix} \begin{bmatrix}420\\0\\120\end{bmatrix} \begin{bmatrix}378\\-42\\126\end{bmatrix}$$

$$B6 = \begin{bmatrix}420\\0\\120\end{bmatrix} \begin{bmatrix}462\\42\\126\end{bmatrix} \begin{bmatrix}420\\0\\140\end{bmatrix} \begin{bmatrix}420\\140\\140\end{bmatrix} \begin{bmatrix}280\\140\\140\end{bmatrix} \begin{bmatrix}210\\210\\210\end{bmatrix}$$

28.7. The Action of the Map

In this section we explain the action of the map on each of the polyhedra listed above. To each polyhedron we associate a 4-tuple of integers. The list $V = (u_1, v_1, u_2, v_2)$ tells us that

$$F_V \begin{bmatrix}x\\y\\s\end{bmatrix} = \begin{bmatrix}1 & 0 & 2v_1 - 2v_2\\0 & 1 & 2v_1 + 2v_2\\0 & 0 & 1\end{bmatrix} \begin{bmatrix}x\\y\\s\end{bmatrix} + \begin{bmatrix}-2u_1\\2u_2\\0\end{bmatrix}. \tag{28.5}$$

Remark: Equation 28.5 gives the equation for the action on the unscaled polyhedra. When we acts on the scaled polyhedra listed above, we need to scale the translation part of the map by 420. That is, $-2u_1$ and $2u_2$ need to be replaced by $-840u_1$ and $840u_2$.

The polyhedra α_0 and β_0 correspond to the trivial tiles. The vectors associated to these are $a_0 = b_0 = (0, 0, 0, 0)$. Below we will list the vectors $a_1, ..., a_9, b_1, ..., b_6$. We have the relations

$$a_{9+i} = -a_i, \qquad b_{6+i} = -b_i. \tag{28.6}$$

The vector c_i associated to $\iota_2 \circ F(\alpha_i)$ is given by the following rule:

$$a_i = (u_1, v_1, u_2, v_2) \implies c_i = (-v_2, -u_2, -v_1, -u_1). \tag{28.7}$$

Recall that the map F is really the composition $(F')^2$, where F' maps the bundle $\mathcal{X}[1/4, 2]$ to the polyhedron obtained by rotating $\mathcal{X}[1/4, 2]$ by 90 degrees about the z-axis. To get the action of F' we simply replace each vector $V = a_1, a_2, ...$ by V', where

$$V = (u_1, v_1, u_2, v_2) \implies V' = (u_1, v_1, 0, 0). \tag{28.8}$$

Here are the vectors.

$$a_1 = (1, 2, 0, -2), \qquad a_2 = (0, -1, -1, -2) \qquad a_3 = (0, -1, -1, -1).$$

$$a_4 = (1, 1, 0, -1), \qquad a_5 = (0, -2, -1, -2) \qquad a_6 = (1, 2, 1, 1).$$

$$a_7 = (0, -2, -1, -1), \qquad a_8 = (0, -1, 0, 1) \qquad a_9 = (1, 2, 1, 2).$$

$$b_1 = (1, 0, -1, -1), \qquad b_2 = (1, 1, 0, -1) \qquad b_3 = (1, 1, 1, 0).$$

$$b_4 = (0, -1, -1, 0), \qquad b_5 = (0, -1, -1, -1) \qquad b_6 = (1, 1, 1, 1).$$

28.8. The Partition for Calculation 11

For the parameter interval $[1/2, 1]$, the parallelogram F_1 is partitioned by the polyhedra $B_0, ..., B_{12}$. For Calculations 11 and 12, it is more convenient to work with a partition of the dogbone D. Recall that $D$is obtained from the domain F_1 by deleting the central tile $F_1 \cap F_2$ and shifting the lower component of $F_1 - F_2$ by the vector $(2, 0)$. See Figure 5.1. Accordingly, we define B'_j to be the translate of B_j which lies in D. Either $B'_j = B_j$ or $B'_j = B_j + (2, 0)$, depending on which component of $F_1 - F_2$ contains B_j.

The partition $\mathcal{A}_2$ mentioned in §27.11 consists of 40 polyhedra $P_0, ..., P_{39}$. Each of these polyhedra has the form

$$P(i, j) = f(B'_i) \cap B'_j. \tag{28.9}$$

To illustrate our notation by way of example, the symbol $P1(10, 2)$ means that the polyhedron $P(10, 2)$ is listed as P_10 in our program. We only list $P_0, ..., P_{19}$. We have

$$P_{j+20} = \rho(P_j), \qquad \rho(x, y, z) = (2 - x, -y, z). \tag{28.10}$$

This time, we scale all the polyhedra by a factor of 840.

$$P0(12, 1) = \begin{bmatrix} 840 \\ -840 \\ 840 \end{bmatrix} \begin{bmatrix} 1680 \\ -840 \\ 840 \end{bmatrix} \begin{bmatrix} 1050 \\ -630 \\ 630 \end{bmatrix} \begin{bmatrix} 1120 \\ -280 \\ 560 \end{bmatrix} \begin{bmatrix} 1050 \\ -210 \\ 630 \end{bmatrix}$$

$$P1(10, 2) = \begin{bmatrix} 1260 \\ -420 \\ 420 \end{bmatrix} \begin{bmatrix} 840 \\ 0 \\ 560 \end{bmatrix} \begin{bmatrix} 1050 \\ 210 \\ 630 \end{bmatrix} \begin{bmatrix} 840 \\ 0 \\ 630 \end{bmatrix} \begin{bmatrix} 1120 \\ -280 \\ 560 \end{bmatrix}$$

$$P2(10, 3) = \begin{bmatrix} 840 \\ 0 \\ 840 \end{bmatrix} \begin{bmatrix} 840 \\ 0 \\ 672 \end{bmatrix} \begin{bmatrix} 1050 \\ -210 \\ 630 \end{bmatrix} \begin{bmatrix} 980 \\ 140 \\ 700 \end{bmatrix}$$

$$P3(12, 4) = \begin{bmatrix} 840 \\ -560 \\ 560 \end{bmatrix} \begin{bmatrix} 630 \\ -630 \\ 630 \end{bmatrix} \begin{bmatrix} 560 \\ -280 \\ 560 \end{bmatrix} \begin{bmatrix} 560 \\ -560 \\ 560 \end{bmatrix} \begin{bmatrix} 525 \\ -315 \\ 525 \end{bmatrix} \begin{bmatrix} 672 \\ -168 \\ 504 \end{bmatrix}$$

$$P4(12, 5) = \begin{bmatrix} 420 \\ -420 \\ 420 \end{bmatrix} \begin{bmatrix} 840 \\ 0 \\ 420 \end{bmatrix} \begin{bmatrix} 840 \\ -560 \\ 560 \end{bmatrix} \begin{bmatrix} 672 \\ -168 \\ 504 \end{bmatrix} \begin{bmatrix} 560 \\ -560 \\ 560 \end{bmatrix} \begin{bmatrix} 525 \\ -315 \\ 525 \end{bmatrix}$$

$$P5(10, 6) = \begin{bmatrix} 1050 \\ 210 \\ 630 \end{bmatrix} \begin{bmatrix} 840 \\ 0 \\ 672 \end{bmatrix} \begin{bmatrix} 1050 \\ -210 \\ 630 \end{bmatrix} \begin{bmatrix} 840 \\ 0 \\ 630 \end{bmatrix} \begin{bmatrix} 1120 \\ -280 \\ 560 \end{bmatrix} \begin{bmatrix} 980 \\ 140 \\ 700 \end{bmatrix}$$

$$P6(12, 6) = \begin{bmatrix} 840 \\ 0 \\ 420 \end{bmatrix} \begin{bmatrix} 840 \\ -560 \\ 560 \end{bmatrix} \begin{bmatrix} 1050 \\ -630 \\ 630 \end{bmatrix} \begin{bmatrix} 840 \\ -840 \\ 840 \end{bmatrix} \begin{bmatrix} 840 \\ 0 \\ 560 \end{bmatrix} \begin{bmatrix} 630 \\ -210 \\ 630 \end{bmatrix} \begin{bmatrix} 630 \\ -630 \\ 630 \end{bmatrix} \begin{bmatrix} 1050 \\ -210 \\ 630 \end{bmatrix} \begin{bmatrix} 1120 \\ -280 \\ 560 \end{bmatrix} \begin{bmatrix} 560 \\ -280 \\ 560 \end{bmatrix}$$

$$P7(7, 7) = \begin{bmatrix} 840 \\ -840 \\ 840 \end{bmatrix} \begin{bmatrix} 0 \\ -840 \\ 840 \end{bmatrix} \begin{bmatrix} 840 \\ 0 \\ 840 \end{bmatrix} \begin{bmatrix} 210 \\ -630 \\ 630 \end{bmatrix}$$

$$P8(8, 7) = \begin{bmatrix} 1680 \\ -840 \\ 840 \end{bmatrix} \begin{bmatrix} 420 \\ -420 \\ 420 \end{bmatrix} \begin{bmatrix} 1120 \\ -280 \\ 560 \end{bmatrix} \begin{bmatrix} 1470 \\ -630 \\ 630 \end{bmatrix}$$

$$P9(8, 8) = \begin{bmatrix} 840 \\ 0 \\ 420 \end{bmatrix} \begin{bmatrix} 420 \\ -420 \\ 420 \end{bmatrix} \begin{bmatrix} 1260 \\ -420 \\ 420 \end{bmatrix} \begin{bmatrix} 1400 \\ -560 \\ 560 \end{bmatrix}$$

$$P10(9,8) = \begin{bmatrix}1680\\-840\\840\end{bmatrix} \begin{bmatrix}840\\0\\560\end{bmatrix} \begin{bmatrix}630\\-210\\630\end{bmatrix} \begin{bmatrix}840\\0\\630\end{bmatrix}$$

$$P11(11,8) = \begin{bmatrix}840\\0\\420\end{bmatrix} \begin{bmatrix}1260\\-420\\420\end{bmatrix} \begin{bmatrix}840\\0\\560\end{bmatrix} \begin{bmatrix}1050\\210\\630\end{bmatrix} \begin{bmatrix}1120\\280\\560\end{bmatrix}$$

$$P12(7,9) = \begin{bmatrix}280\\-560\\560\end{bmatrix} \begin{bmatrix}210\\-630\\630\end{bmatrix} \begin{bmatrix}840\\0\\840\end{bmatrix} \begin{bmatrix}840\\-840\\840\end{bmatrix} \begin{bmatrix}630\\-630\\630\end{bmatrix}$$

$$P13(9,9) = \begin{bmatrix}840\\0\\840\end{bmatrix} \begin{bmatrix}840\\-840\\840\end{bmatrix} \begin{bmatrix}1680\\-840\\840\end{bmatrix} \begin{bmatrix}840\\0\\672\end{bmatrix} \begin{bmatrix}700\\-140\\700\end{bmatrix}$$

$$P14(7,10) = \begin{bmatrix}280\\-560\\560\end{bmatrix} \begin{bmatrix}630\\-210\\630\end{bmatrix} \begin{bmatrix}840\\0\\840\end{bmatrix} \begin{bmatrix}630\\-630\\630\end{bmatrix} \begin{bmatrix}560\\-560\\560\end{bmatrix}$$

$$P15(11,10) = \begin{bmatrix}1260\\-420\\420\end{bmatrix} \begin{bmatrix}1120\\280\\560\end{bmatrix} \begin{bmatrix}1155\\315\\525\end{bmatrix} \begin{bmatrix}1008\\168\\504\end{bmatrix}$$

$$P16(7,11) = \begin{bmatrix}420\\-420\\420\end{bmatrix} \begin{bmatrix}280\\-560\\560\end{bmatrix} \begin{bmatrix}630\\-210\\630\end{bmatrix} \begin{bmatrix}560\\-560\\560\end{bmatrix}$$

$$P17(11,11) = \begin{bmatrix}1260\\-420\\420\end{bmatrix} \begin{bmatrix}1260\\420\\420\end{bmatrix} \begin{bmatrix}840\\0\\420\end{bmatrix} \begin{bmatrix}1008\\168\\504\end{bmatrix} \begin{bmatrix}1155\\315\\525\end{bmatrix}$$

$$P18(8,12) = \begin{bmatrix}840\\0\\420\end{bmatrix} \begin{bmatrix}420\\-420\\420\end{bmatrix} \begin{bmatrix}1470\\-630\\630\end{bmatrix} \begin{bmatrix}1400\\-560\\560\end{bmatrix} \begin{bmatrix}1120\\-280\\560\end{bmatrix}$$

$$P19(9,12) = \begin{bmatrix}840\\-840\\840\end{bmatrix} \begin{bmatrix}1680\\-840\\840\end{bmatrix} \begin{bmatrix}630\\-210\\630\end{bmatrix} \begin{bmatrix}840\\0\\672\end{bmatrix} \begin{bmatrix}840\\0\\630\end{bmatrix} \begin{bmatrix}700\\-140\\700\end{bmatrix}$$

28.9. The First Partition for Calculation 12

The calculation $\mathcal{A}_3$ is obtained by adding another 40 polyhedra to the 40 already contained in $\mathcal{A}_2$. Let ρ be the map from Equation 28.10. Here are the equations for the maps discussed in §27.12.

- $\mathbf{up}(x,y,z) = (x+2z-2, y+2z, z)$.
- $\mathbf{dn}(x,y,z) = (x+2z-2, y-2z, z)$

The pieces $P_{40}, ..., P_{59}$ have the form

$$P_{40+j} = \mathbf{up}(P_j), \qquad j = 0, ..., 19. \tag{28.11}$$

The pieces $P_{60}, ..., P_{79}$ have a more *ad hoc* description. As in Equation 28.10, let $\rho(x,y,z) = (2-x, -y, z)$. It is tempting to define

$$P_{60+j} = \mathbf{dn} \circ \rho(P_j),$$

but this gives us pieces which cover the wrong region. To get exactly the right pieces, we define

- $\rho'(p) = \rho(p)$ unless $p \in P_j$ for $j \in 1, 2, 5, 11, 15, 17$.
- $\rho(p) = p$ when $p \in P_j$ for $j \in 1, 2, 5, 11, 15, 17$.

Then we define

$$P_{60+j} = \mathbf{dn} \circ \rho'(P_j), \qquad j = 0, ..., 19. \tag{28.12}$$

This gives us a covering of the bottom component of $E_1 - D$, as desired.

28.10. The Second Partition for Calculation 12

The partition $\mathcal{A}_4$ uses 76 polyhedra from $\mathcal{A}_3$, and replaces the 4 missing polyhedra with 25 new ones. The omitted 4 polyhedra are

$$P_j, \qquad j = 0, 28, 58, 67. \tag{28.13}$$

Here are the 25 new polyhedra. The scale factor is $83160 = 840 \times 99$.

$$Q0 = \begin{bmatrix} 103950 \\ -62370 \\ 62370 \end{bmatrix} \begin{bmatrix} 124740 \\ -41580 \\ 62370 \end{bmatrix} \begin{bmatrix} 118800 \\ -35640 \\ 59400 \end{bmatrix} \begin{bmatrix} 106920 \\ -23760 \\ 59400 \end{bmatrix} \begin{bmatrix} 108108 \\ -24948 \\ 58212 \end{bmatrix}$$

$$Q1 = \begin{bmatrix} 103950 \\ -20790 \\ 62370 \end{bmatrix} \begin{bmatrix} 116424 \\ -66528 \\ 66528 \end{bmatrix} \begin{bmatrix} 99792 \\ -66528 \\ 66528 \end{bmatrix} \begin{bmatrix} 97020 \\ -69300 \\ 69300 \end{bmatrix}$$

$$Q2 = \begin{bmatrix} 103950 \\ -62370 \\ 62370 \end{bmatrix} \begin{bmatrix} 103950 \\ -20790 \\ 62370 \end{bmatrix} \begin{bmatrix} 116424 \\ -66528 \\ 66528 \end{bmatrix} \begin{bmatrix} 124740 \\ -41580 \\ 62370 \end{bmatrix} \begin{bmatrix} 106920 \\ -23760 \\ 59400 \end{bmatrix} \begin{bmatrix} 99792 \\ -66528 \\ 66528 \end{bmatrix}$$

$$Q3 = \begin{bmatrix} 83160 \\ -83160 \\ 83160 \end{bmatrix} \begin{bmatrix} 166320 \\ -83160 \\ 83160 \end{bmatrix} \begin{bmatrix} 142560 \\ -59400 \\ 71280 \end{bmatrix} \begin{bmatrix} 135135 \\ -51975 \\ 72765 \end{bmatrix} \begin{bmatrix} 135135 \\ -72765 \\ 72765 \end{bmatrix}$$

$$Q4 = \begin{bmatrix} 103950 \\ -20790 \\ 62370 \end{bmatrix} \begin{bmatrix} 124740 \\ -41580 \\ 62370 \end{bmatrix} \begin{bmatrix} 116424 \\ -33264 \\ 66528 \end{bmatrix} \begin{bmatrix} 97020 \\ -41580 \\ 69300 \end{bmatrix} \begin{bmatrix} 99792 \\ -49896 \\ 66528 \end{bmatrix}$$

$$Q5 = \begin{bmatrix} 103950 \\ -62370 \\ 62370 \end{bmatrix} \begin{bmatrix} 110880 \\ -27720 \\ 55440 \end{bmatrix} \begin{bmatrix} 118800 \\ -35640 \\ 59400 \end{bmatrix} \begin{bmatrix} 108108 \\ -24948 \\ 58212 \end{bmatrix}$$

$$Q6 = \begin{bmatrix} 83160 \\ -83160 \\ 83160 \end{bmatrix} \begin{bmatrix} 142560 \\ -59400 \\ 71280 \end{bmatrix} \begin{bmatrix} 138600 \\ -55440 \\ 69300 \end{bmatrix} \begin{bmatrix} 124740 \\ -41580 \\ 69300 \end{bmatrix} \begin{bmatrix} 135135 \\ -51975 \\ 72765 \end{bmatrix} \begin{bmatrix} 135135 \\ -72765 \\ 72765 \end{bmatrix} \begin{bmatrix} 130680 \\ -71280 \\ 71280 \end{bmatrix} \begin{bmatrix} 114345 \\ -72765 \\ 72765 \end{bmatrix}$$

$$Q7 = \begin{bmatrix} 133056 \\ -49896 \\ 66528 \end{bmatrix} \begin{bmatrix} 138600 \\ -55440 \\ 69300 \end{bmatrix} \begin{bmatrix} 124740 \\ -41580 \\ 69300 \end{bmatrix} \begin{bmatrix} 124740 \\ -69300 \\ 69300 \end{bmatrix} \begin{bmatrix} 130680 \\ -71280 \\ 71280 \end{bmatrix} \begin{bmatrix} 114345 \\ -72765 \\ 72765 \end{bmatrix}$$

$$Q8 = \begin{bmatrix} 83160 \\ -83160 \\ 83160 \end{bmatrix} \begin{bmatrix} 116424 \\ -66528 \\ 66528 \end{bmatrix} \begin{bmatrix} 124740 \\ -41580 \\ 62370 \end{bmatrix} \begin{bmatrix} 124740 \\ -69300 \\ 69300 \end{bmatrix} \begin{bmatrix} 133056 \\ -49896 \\ 66528 \end{bmatrix} \begin{bmatrix} 116424 \\ -33264 \\ 66528 \end{bmatrix} \begin{bmatrix} 124740 \\ -41580 \\ 69300 \end{bmatrix} \begin{bmatrix} 97020 \\ -41580 \\ 69300 \end{bmatrix} \begin{bmatrix} 97020 \\ -69300 \\ 69300 \end{bmatrix} \begin{bmatrix} 99792 \\ -49896 \\ 66528 \end{bmatrix}$$

$$Q9 = \begin{bmatrix} 83160 \\ -166320 \\ 83160 \end{bmatrix} \begin{bmatrix} 116424 \\ -66528 \\ 66528 \end{bmatrix} \begin{bmatrix} 97020 \\ -97020 \\ 69300 \end{bmatrix} \begin{bmatrix} 99792 \\ -66528 \\ 66528 \end{bmatrix} \begin{bmatrix} 97020 \\ -69300 \\ 69300 \end{bmatrix}$$

$$Q10 = \begin{bmatrix} 103950 \\ -62370 \\ 62370 \end{bmatrix} \begin{bmatrix} 116424 \\ -66528 \\ 66528 \end{bmatrix} \begin{bmatrix} 97020 \\ -97020 \\ 69300 \end{bmatrix} \begin{bmatrix} 99792 \\ -66528 \\ 66528 \end{bmatrix}$$

$$Q11 = \begin{bmatrix} 83160 \\ -83160 \\ 83160 \end{bmatrix} \begin{bmatrix} 166320 \\ -83160 \\ 83160 \end{bmatrix} \begin{bmatrix} 83160 \\ -166320 \\ 83160 \end{bmatrix} \begin{bmatrix} 135135 \\ -72765 \\ 72765 \end{bmatrix}$$

$$Q12 = \begin{bmatrix} 83160 \\ -83160 \\ 83160 \end{bmatrix} \begin{bmatrix} 83160 \\ -166320 \\ 83160 \end{bmatrix} \begin{bmatrix} 135135 \\ -72765 \\ 72765 \end{bmatrix} \begin{bmatrix} 130680 \\ -71280 \\ 71280 \end{bmatrix} \begin{bmatrix} 114345 \\ -72765 \\ 72765 \end{bmatrix}$$

$$Q13 = \begin{bmatrix} 83160 \\ -83160 \\ 83160 \end{bmatrix} \begin{bmatrix} 83160 \\ -166320 \\ 83160 \end{bmatrix} \begin{bmatrix} 116424 \\ -66528 \\ 66528 \end{bmatrix} \begin{bmatrix} 124740 \\ -69300 \\ 69300 \end{bmatrix} \begin{bmatrix} 97020 \\ -69300 \\ 69300 \end{bmatrix}$$

$$Q14 = \begin{bmatrix} 83160 \\ -166320 \\ 83160 \end{bmatrix} \begin{bmatrix} 124740 \\ -69300 \\ 69300 \end{bmatrix} \begin{bmatrix} 130680 \\ -71280 \\ 71280 \end{bmatrix} \begin{bmatrix} 114345 \\ -72765 \\ 72765 \end{bmatrix}$$

$$Q15 = \begin{bmatrix}83160\\55440\\55440\end{bmatrix} \begin{bmatrix}41580\\41580\\62370\end{bmatrix} \begin{bmatrix}35640\\47520\\59400\end{bmatrix} \begin{bmatrix}41580\\41580\\58212\end{bmatrix} \begin{bmatrix}41580\\58212\\58212\end{bmatrix} \begin{bmatrix}55440\\55440\\55440\end{bmatrix} \begin{bmatrix}47520\\35640\\59400\end{bmatrix} \begin{bmatrix}47520\\59400\\59400\end{bmatrix}$$

$$Q16 = \begin{bmatrix}20790\\62370\\62370\end{bmatrix} \begin{bmatrix}41580\\41580\\62370\end{bmatrix} \begin{bmatrix}35640\\47520\\59400\end{bmatrix} \begin{bmatrix}41580\\58212\\58212\end{bmatrix} \begin{bmatrix}47520\\59400\\59400\end{bmatrix}$$

$$Q17 = \begin{bmatrix}0\\83160\\83160\end{bmatrix} \begin{bmatrix}10395\\72765\\72765\end{bmatrix} \begin{bmatrix}23760\\59400\\71280\end{bmatrix} \begin{bmatrix}41580\\69300\\69300\end{bmatrix}$$

$$Q18 = \begin{bmatrix}20790\\62370\\62370\end{bmatrix} \begin{bmatrix}83160\\55440\\55440\end{bmatrix} \begin{bmatrix}62370\\62370\\62370\end{bmatrix} \begin{bmatrix}41580\\41580\\62370\end{bmatrix} \begin{bmatrix}16632\\66528\\66528\end{bmatrix}$$

$$Q19 = \begin{bmatrix}55440\\27720\\55440\end{bmatrix} \begin{bmatrix}83160\\55440\\55440\end{bmatrix} \begin{bmatrix}41580\\41580\\58212\end{bmatrix} \begin{bmatrix}55440\\55440\\55440\end{bmatrix} \begin{bmatrix}47520\\35640\\59400\end{bmatrix}$$

$$Q20 = \begin{bmatrix}13860\\69300\\69300\end{bmatrix} \begin{bmatrix}10395\\72765\\72765\end{bmatrix} \begin{bmatrix}23760\\59400\\71280\end{bmatrix} \begin{bmatrix}27720\\55440\\69300\end{bmatrix} \begin{bmatrix}41580\\69300\\69300\end{bmatrix} \begin{bmatrix}49896\\66528\\66528\end{bmatrix}$$

$$Q21 = \begin{bmatrix}62370\\62370\\62370\end{bmatrix} \begin{bmatrix}13860\\69300\\69300\end{bmatrix} \begin{bmatrix}33264\\49896\\66528\end{bmatrix} \begin{bmatrix}27720\\55440\\69300\end{bmatrix} \begin{bmatrix}49896\\66528\\66528\end{bmatrix}$$

$$Q22 = \begin{bmatrix}62370\\62370\\62370\end{bmatrix} \begin{bmatrix}41580\\41580\\62370\end{bmatrix} \begin{bmatrix}16632\\66528\\66528\end{bmatrix} \begin{bmatrix}13860\\69300\\69300\end{bmatrix} \begin{bmatrix}33264\\49896\\66528\end{bmatrix}$$

$$Q23 = \begin{bmatrix}83160\\55440\\55440\end{bmatrix} \begin{bmatrix}55440\\83160\\55440\end{bmatrix} \begin{bmatrix}74844\\58212\\58212\end{bmatrix} \begin{bmatrix}55440\\55440\\55440\end{bmatrix} \begin{bmatrix}95040\\59400\\59400\end{bmatrix}$$

$$Q24 = \begin{bmatrix}103950\\62370\\62370\end{bmatrix} \begin{bmatrix}55440\\83160\\55440\end{bmatrix} \begin{bmatrix}74844\\58212\\58212\end{bmatrix} \begin{bmatrix}95040\\59400\\59400\end{bmatrix}$$

Bibliography

[AE] Shigeki Akiyama and Edmund Harris, *Pentagonal Domain Exchange*, preprint, 2012

[AG] A. Goetz and G. Poggiaspalla, *Rotations by $\pi/7$*, Nonlinearity **17** (2004) no. 5 1787-1802

[AKT] R. Adler, B. Kitchens, and C. Tresser, *Dynamics of non-ergodic piecewise affine maps of the torus*, Ergodic Theory Dyn. Syst **21** (2001) no. 4 959-999

[BC] N. Bedaride and J. Cassaigne, *Outer Billiards outside regular polygons*, J. London Math Soc. (2011)

[BKS] T. Bedford, M. Keane, and C. Series, eds., *Ergodic Theory, Symbolic Dynamics, and Hyperbolic Spaces*, Oxford University Press, Oxford (1991).

[Be] A. Beardon, *The Geometry of Discrete Groups*, Graduate Texts in Mathematics 91, Springer, New York (1983).

[Bo] P. Boyland, *Dual billiards, twist maps, and impact oscillators*, Nonlinearity **9**:1411–1438 (1996).

[B] J, Buzzi, *Piecewise isometrries have zero topological entropy (English summary)* Ergodic Theory and Dynamical Systems **21** (2001) no. 5 pp 1371-1377

[Da] Davenport, *The Higher Arithmetic: An Introduction to the Theory of Numbers*, Hutchinson and Company, London (1952).

[DT1] F. Dogru and S. Tabachnikov, *Dual billiards*, Math. Intelligencer **26**(4):18–25 (2005).

[DT2] F. Dogru and S. Tabachnikov, *Dual billiards in the hyperbolic plane*, Nonlinearity **15**:1051–1072 (2003).

[F] K. J. Falconer, *Fractal Geometry: Mathematical Foundations and Applications*, John Wiley and Sons, New York (1990).

[G] D. Genin, *Regular and Chaotic Dynamics of Outer Billiards*, Pennsylvania State University Ph.D. thesis, State College (2005).

[Go] A. Goetz, *Piecewise Isometries – an emerging area of dynamical systems*, preprint.

[GH1] E Gutkin and N. Haydn, *Topological entropy of generalized polygon exchanges*, Bull. Amer. Math. Soc., **32** (1995) no. 1., pp 50-56

[GH2] E Gutkin and N. Haydn, *Topological entropy polygon exchange transformations and polygonal billiards*, Ergodic Theory and Dynamical Systems **17** (1997) no. 4., pp 849-867

[GS] E. Gutkin and N. Simanyi, *Dual polygonal billiard and necklace dynamics*, Comm. Math. Phys. **143**:431–450 (1991).

[H] H. Haller, *Rectangle Exchange Transformations*, Monatsh Math. **91** (1985) 215-232

[Hoo] W. Patrick Hooper, *Renormalization of Polygon Exchage Maps arising from Corner Percolation* Invent. Math. 2012.

[K] M. Keane, *Non-Ergodic Interval Exchange Transformations*, Israel Journal of Math, **26**, 188-96 (1977)

[LKV] J. H. Lowenstein, K. L. Koupsov, F. Vivaldi, *Recursive Tiling and Geometry of piecewise rotations by $\pi/7$*, nonlinearity **17** (2004) no. 2.

[Low1] J. H. Lowenstein, *Aperiodic orbits of piecewise rational rotations of convex polygons with recursive tiling*, Dyn. Syst. **22** (2007) no. 1 25-63

[Low2] J. H. Lowenstein, *Pseudochaotic kicked oscillators*, Springer (2012)

[Ko] Kolodziej, *The antibilliard outside a polygon*, Bull. Pol. Acad Sci. Math. **37**:163–168 (1994).

[N] B. H. Neumann, *Sharing ham and eggs*, Summary of a Manchester Mathematics Colloquium, 25 Jan 1959, published in Iota, the Manchester University Mathematics Students' Journal.

[R] G. Rauzy, *Exchanges d'intervalles et transformations induites*, Acta. Arith. **34** 315-328 (1979)

[S1] R. E. Schwartz, *Mostly Surfaces*, A.M.S. Student Lecture Series **50** 2011.

[S2] R.E. Schwartz *Outer Billiards, Quarter Turn Compositions, and Polytope Exchange Transformations*, preprint (2011)

[S3] R. E. Schwartz, *Outer Billiards on Kites*, Annals of Math Studies **171**, Princeton University Press (2009)

[S4] R. E. Schwartz, *Outer Billiards on the Penrose Kite: Compactification and Renormalization*, Journal of Modern Dynamics, 2012.

[S5] R. E. Schwartz, *Outer Billirds, the Octagon, and the Arithmetic Graph*, preprint (2010)

[S6] R. E. Schwartz, *Square Turning Maps and their Compactifications*, preprint (2013)

[T1] S. Tabachnikov, *Geometry and billiards*, Student Mathematical Library 30, Amer. Math. Soc. (2005).

[T2] S. Tabachnikov, *Billiards*, Société Mathématique de France, "Panoramas et Syntheses" 1, 1995

[VL] F. Vivaldi and J. H. Lowenstein, —it Arithmetical properties of a family of irrational piecewise rotations, *Nonlinearity* **19**:1069–1097 (2007).

[VS] F. Vivaldi and A. Shaidenko, *Global stability of a class of discontinuous dual billiards*, Comm. Math. Phys. **110**:625–640 (1987).

[W] S. Wolfram, *The Mathematica Book*, 4th ed., Wolfram Media/Cambridge University Press, Champaign/Cambridge (1999).

[Y] J.-C. Yoccoz, *Continued Fraction Algorithms for Interval Exchange Maps: An Introduction* , Frontiers in Number Theory, Physics, and Geometry Vol 1, P. Cartier, B. Julia, P. Moussa, P. Vanhove (ed.) Springer-Verlag 4030437 (2006)

[Z] A. Zorich, *Flat Surfaces*, Frontiers in Number Theory, Physics, and Geometry Vol 1, P. Cartier, B. Julia, P. Moussa, P. Vanhove (editors) Springer-Verlag 4030437 (2006)

Selected Published Titles in This Series

197 **Richard Evan Schwartz,** The Octogonal PETs, 2014

196 **Silouanos Brazitikos, Apostolos Giannopoulos, Petros Valettas, and Beatrice-Helen Vritsiou,** Geometry of Isotropic Convex Bodies, 2014

195 **Ching-Li Chai, Brian Conrad, and Frans Oort,** Complex Multiplication and Lifting Problems, 2013

194 **Samuel Herrmann, Peter Imkeller, Ilya Pavlyukevich, and Dierk Peithmann,** Stochastic Resonance, 2014

193 **Robert Rumely,** Capacity Theory with Local Rationality, 2013

192 **Messoud Efendiev,** Attractors for Degenerate Parabolic Type Equations, 2013

191 **Grégory Berhuy and Frédérique Oggier,** An Introduction to Central Simple Algebras and Their Applications to Wireless Communication, 2013

190 **Aleksandr Pukhlikov,** Birationally Rigid Varieties, 2013

189 **Alberto Elduque and Mikhail Kochetov,** Gradings on Simple Lie Algebras, 2013

188 **David Lannes,** The Water Waves Problem, 2013

187 **Nassif Ghoussoub and Amir Moradifam,** Functional Inequalities: New Perspectives and New Applications, 2013

186 **Gregory Berkolaiko and Peter Kuchment,** Introduction to Quantum Graphs, 2013

185 **Patrick Iglesias-Zemmour,** Diffeology, 2013

184 **Frederick W. Gehring and Kari Hag,** The Ubiquitous Quasidisk, 2012

183 **Gershon Kresin and Vladimir Maz'ya,** Maximum Principles and Sharp Constants for Solutions of Elliptic and Parabolic Systems, 2012

182 **Neil A. Watson,** Introduction to Heat Potential Theory, 2012

181 **Graham J. Leuschke and Roger Wiegand,** Cohen-Macaulay Representations, 2012

180 **Martin W. Liebeck and Gary M. Seitz,** Unipotent and Nilpotent Classes in Simple Algebraic Groups and Lie Algebras, 2012

179 **Stephen D. Smith,** Subgroup complexes, 2011

178 **Helmut Brass and Knut Petras,** Quadrature Theory, 2011

177 **Alexei Myasnikov, Vladimir Shpilrain, and Alexander Ushakov,** Non-commutative Cryptography and Complexity of Group-theoretic Problems, 2011

176 **Peter E. Kloeden and Martin Rasmussen,** Nonautonomous Dynamical Systems, 2011

175 **Warwick de Launey and Dane Flannery,** Algebraic Design Theory, 2011

174 **Lawrence S. Levy and J. Chris Robson,** Hereditary Noetherian Prime Rings and Idealizers, 2011

173 **Sariel Har-Peled,** Geometric Approximation Algorithms, 2011

172 **Michael Aschbacher, Richard Lyons, Stephen D. Smith, and Ronald Solomon,** The Classification of Finite Simple Groups, 2011

171 **Leonid Pastur and Mariya Shcherbina,** Eigenvalue Distribution of Large Random Matrices, 2011

170 **Kevin Costello,** Renormalization and Effective Field Theory, 2011

169 **Robert R. Bruner and J. P. C. Greenlees,** Connective Real K-Theory of Finite Groups, 2010

168 **Michiel Hazewinkel, Nadiya Gubareni, and V. V. Kirichenko,** Algebras, Rings and Modules, 2010

167 **Michael Gekhtman, Michael Shapiro, and Alek Vainshtein,** Cluster Algebras and Poisson Geometry, 2010

166 **Kyung Bai Lee and Frank Raymond,** Seifert Fiberings, 2010

165 **Fuensanta Andreu-Vaillo, José M. Mazón, Julio D. Rossi, and J. Julián Toledo-Melero,** Nonlocal Diffusion Problems, 2010

For a complete list of titles in this series, visit the
AMS Bookstore at **www.ams.org/bookstore/survseries/**.